THE ASSESSMENT AND PERCEPTION OF RISK

THE ASSESSMENT AND PERCEPTION OF RISK

A ROYAL SOCIETY DISCUSSION

ORGANIZED BY
SIR FREDERICK WARNER, F.R.S.,
AND D. H. SLATER,
ON BEHALF OF
THE ROYAL SOCIETY'S STUDY GROUP
ON RISK

HELD ON 12 AND 13 NOVEMBER 1980

LONDON
THE ROYAL SOCIETY
1981

Printed in Great Britain for the Royal Society
at the
University Press, Cambridge

ISBN 0 85403 163 4

First published in *Proceedings of the Royal Society of London*,
series A, volume 376 (no. 1764), pages 1–206

Published by the Royal Society
6 Carlton House Terrace, London SW1Y 5AG

PREFACE

At the end of 1978, the Royal Society set up a Study Group on Risk. This has met at intervals to discuss the ways in which risks can be measured and also the ways in which they are perceived. As part of the Group's work, it was decided to invite a number of specialists to present papers at a two-day Discussion Meeting. This was held at the Society in November 1980.

The techniques of risk assessment go back to the need for reliability in military equipment and air transport. They can now use failure rates held in data banks containing rapidly increasing amounts of information. The use of fault-trees in risk analysis gives a logical basis for reducing risk during the conceptual stage of projects. A number of papers deal with the problems of assessment shown up in conventional, unconventional, and nuclear power plants. The problems have existed much longer in the risks of failure in bridges, dams, buildings, chemical and petroleum installations, and transport. Risks associated with drugs and medical procedures are complicated by the benefits weighed against them; the risks also show up only over long periods as a result of epidemological studies and finally in mortality tables. The papers discuss not only the risk that is final, that of death, but also of injury up to fates worse than death, with special reference to risks at work.

A concluding section of the papers deals with the economist's view of cost–benefit analysis as comprehending assessment and perception, the industry view of risk-taking and the operation of agencies charged with regulating risks.

This volume, which I have edited, should contribute to the debate about using resources economically, effectively and with proper regard to the benefits that people derive from their use and their fears about the degree and kind of risks involved.

February 1980 FREDERICK WARNER

CONTENTS

Proc. R. Soc. Lond. A **376**, 3–4 (1981)
Printed in Great Britain

Introductory remarks

By Lord Ashby, F.R.S.

In 1947 the distinguished American jurist, Judge Learned Hand, had to deal with a case of negligence against the owners of a tugboat that had caused a barge to sink. His judgment in the case included a terse summary of the issues that we are here to discuss. This is what he said:

> The owner's duty...to provide against resulting injuries is a function of three variables: (1) the probability that [the tugboat] will break away; (2) the gravity of the resulting injury, if she does; (3) the burden of adequate precautions [to prevent her breaking away]...If the probability be called P; the injury L; and the burden B; liability depends upon whether B is less than L multiplied by P.

Decision-makers in public affairs, industry and professions like architecture, engineering and medicine, are expected to make estimates of P, L and B. They all encounter two formidable difficulties. One is to get reliable quantitative values for P, L and B. The other difficulty is to know what weight is attached to these values (if they can be got) by the people on whose behalf the decisions are being made.

Despite a great deal of work, neither of these difficulties has yet been resolved. The desire for precision in risk–benefit analysis tempts people to quantify values in ways that do not generate much confidence. Thus the cost–benefit analysis used in the design of trunk roads includes an estimate of the value to be attached to a fatal casualty. This (in 1976) was about £42000, of which £13000 was 'an allowance for pain, grief, and suffering.' The comparable value for Canadians was £66000 and for Australians £78000! It is doubtful whether this sort of quantification helps those who have to make decisions about risk-management.

The second difficulty is to know how people perceive the statistical data for frequency and severity of misfortunes. This is even more perplexing for decision makers because many political decisions have to attach more weight to the perception of a risk than to its statistical assessment. This is true for the political decision in Britain to lower the level of lead additives to petrol. The strength of this subjective perception has also been summarized tersely, by a Fellow of this Society, Sir Harold Wilson, when he was Prime Minister. He was speaking about unemployment, which was at that time running at about 7%. But, he said, for the man who is unemployed, the level is 100%.

It will not do to dismiss as irrational this disparity between assessments of risk (of unemployment and many other things) and perceptions of those risks by people exposed to them. And it will not do to suppose that the disparity can be resolved by education. The challenge to the decision-maker is to create an acceptable policy in the light of this disparity. One class of experts has successfully met this challenge, namely underwriters who deal with insurance. For them to remain solvent their calculations have to be based on assessed risk; otherwise they would go out of business. But to attract customers their advertisements must make some appeal to perceived risk; otherwise they would get no business.

Thus both lawyers and underwriters have useful contributions to make to the theme of this meeting and I invite any who are here to take part.

We begin this morning with the second of the two difficulties that I mentioned: the perception of risk.

Proc. R. Soc. Lond. A **376**, 5–16 (1981)
Printed in Great Britain

PERCEPTION OF RISK

The public's perception of risk and the question of irrationality

By T. R. Lee
Department of Psychology, University of Surrey, Guildford, Surrey GU2 5XH, U.K.

The future likelihood of a hazardous event is predicted from its frequency per unit of past time or by aggregating the influence of the variables that determine it. It is assumed that accurate scientific measurement provides the best estimate, and that public perceptions, if discrepant, are 'irrational'.

Certainly, perception is a constructive process, dependent on partial information, selected again at the receptor stage and distorted in memory store. However, so-called 'objective' assessments generally subsume a limited range of variables, discount interactions and are biased to variables that are readily quantifiable for recording or future modelling. Extrapolation from the past rashly assumes stable continuity.

In assessing 'seriousness' as distinct from probability, it may be argued that only the public can judge, since social and moral values are the ultimate criteria.

Clearly, both objective and perceived assessments must be considered and eventually reconciled. The problems of measurement and a quest for lawfulness in the perception of risk are discussed.

Introduction and definition

A *hazard* is a situation or activity involving events whose consequences are undesirable to some unknown degree and whose future occurrence is uncertain.

Our everyday perceptions of hazards usually compound the probability that accidental events will occur with their likely damage to property or loss of life if they should occur. These together constitute a general dimension of *risk* from negligible to severe. It is not uncommon, however, for us to be more differentiated in our evaluation, representing a variety of possible events associated with the hazard and attaching different predictions to their likely occurrence.

The assessment of risk can hardly be regarded as a new form of reasoning for man, and indeed we appear to have evolved, in common with other primates and with mammals, a complex and sensitive biological mechanism that generates motivation for the coping actions that are a natural consequence of our risk assessments. This mechanism is called *fear*, or, in its milder manifestations, anxiety. When we experience this emotion, we devise actions to subdue it. At the societal level this is now called 'risk management'.

Because of our rapidly increasing power to control the environment through science and technology, the assessment of hazards is becoming highly sophisticated, a matter for specialists.

Another contributory factor which should not be underestimated is the profound philosophical change that we have undergone in Western society. Fatalism is discredited and religion is changing. Following the example set by scientists, ordinary people look for the causes of catastrophe in antecedent events rather than in some predetermined fate or in the will of God. Responsibility, if not omniscience, has been transferred to the government and to scientists or other experts. As a reflection of these trends, two related interdisciplinary fields of scientific study have emerged.

Objective risk assessment

The first, which I shall call 'objective' risk assessment, is concerned mainly with the prediction of future events from statistical data provided by past events. It employs two forms of extrapolation, one based on the frequency of past occurrences of the event itself and the other based on an aggregation of frequencies for a whole set of antecedent happenings that are thought to determine the event. The latter approach is the only possible objective one for predicting events for which we have no past records, such as a core melt-down or a collision between two Boeing 747s.

It can be argued that the task of science is mainly to improve and extend these measurement techniques and to make them available for the better government of the people and the more efficient and humane management of industry.

It is exceedingly evident, however, that the public's evaluation of many well known risks is markedly different from the objective assessments made of the same risks by scientists; this brings us to the second, though related, field of risk assessment: the *perception* of risk.

This will form the main subject of my paper, but I should like to preface this by some comparisons between the two approaches.

Risk perception compared with objective risk assessment

Ordinary people form their own assessments of risk. These are personally and individually constructed from unique sensory experiences. Nevertheless, because they share common experiences and also because they attach considerable credibility to information and to assessments provided by others, there is a convergence towards norms of risk perception.

It is tempting but false to think of objective assessment as a moderately accurate way of arriving at the actual risk and of perceptions as dealing only with 'imaginary' or 'irrational' risks, a process subject to bias or error.

Such an argument can hardly be sustained even logically, because whatever can be done to establish the probability of occurrence by extrapolating from the past, there is no way in which the adverse effects of a hazardous event can be fully evaluated except in terms of human values and emotions. A hazard has no meaning except in human terms. For example, it is often presumed that because the preservation of life is an important human value, a simple estimate of the predicted number of deaths should provide a valid index of risk on a linear scale of severity. But as Green (1980) has shown, there are 'fates worse than death' in the public's view and, as Slovic (1979) has shown, considerable *dread* is associated with hazards that nevertheless have very low death rates.

One of several reasons why objective assessments of risks differ so much from public perceptions is that the necessary statistics are simply not available to extrapolate from the *particular* past events that the public regards as important. It is obvious, for example, that morbidity statistics are less readily available than those for mortality, but it is less obvious that death is a simple binary event – you are either dead or you are not – whereas suffering lies on a continuous scale of severity, for which there are very few data available.

We can extend this argument further, by pointing out that the criteria used by the public in assessing the seriousness of an event often include not merely biological states but ethical and moral convictions. The awfulness of a catastrophe may lie not merely in the loss of life and suffering involved, but in the violation of people's sense of justice or moral rightness. These are indeed intangible matters and they change with time.

Another difficulty is that in calculating the severity as distinct from the probability of a hazardous event, a great deal depends on *who* is affected. Each of the significant groupings who are unequally afflicted by the event will have established differing assessments of its seriousness.

This is by no means a new problem for the politician, who has always had to consider trade-offs between one group of people and another. It contributes little to regard one group as more 'rational' than another.

Of course, those scientists who have been concerned with developing the field of objective risk assessment have responded promptly to reduce these shortcomings. What began as a relatively simple operation, in which a representative cross section of risks affecting the public as a whole were placed in order on a scale and divided by some cut-offs into 'acceptable' and 'unacceptable' risks, is giving way to very much more sophisticated approaches in which subsets of risks, where like can more legitimately be compared with like, are being set up and, beginning with Starr's (1969) classic paper, attempts are being made to incorporate considerations of benefit as well as of disbenefit. Complex techniques, such as cost–benefit analysis, cost-effectiveness and risk balancing are being applied and presumably approximating more closely to public perceptions as they include more variables.

This is obviously a desirable trend, but we have to be aware that gaining effective measures of intangible *benefits* is, if anything, even more difficult than assessing costs. Also, the aggregation of mean benefits and costs across a population is a long way from aggregating the risk assessments formed from these costs and benefits by each individual.

The most important point that I wish to emphasize, though, is that these efforts on the part of scientists at the objective end of the continuum are, in effect, attempting to incorporate subjective perceptions and hence to narrow the gap between the two approaches.

The construction of risk perceptions

Let me turn now to the subjective side of things. The closing of the gap is also evident here. The study of perception and cognition is concerned with the ways in which the individual comes to know his environment through the data that impinge on his sense organs. The most distinctive feature of man's central nervous system is its enormous capacity for storing and structuring the residues of past perceptions and bringing these to bear on the processing of fresh input. Much of the contemporary research in cognitive psychology operates with a model of man that is not a 'passive recorder' but an active, seeking, *constructor* of an internal representation of the environment that will enable him to satisfy his developing needs, to *cope* with the pervasive problem of adapting to the environment. As Bartlett (1932) put it, man makes a continual 'effort after meaning'.

Organization into cognitive structures is the main theme of this effort and it is carried out, on the one hand, by selectively assimilating information that is compatible with the existing structures and on the other by adapting these structures so that information can more readily be assimilated. Perception is the process of endowing sensation with meaning, and we do this by dividing and labelling our constructions of reality and by developing unique dimensions along which reality can be construed.

Consider, for example, the perception of an event such as death by fire or from lung cancer. We have to form cognitive schemata that lie on dimensions of 'whatness', 'whereness' and 'whenness'; that is, of the nature, the location and the timing of the event.

We assimilate our own direct experiences, but these are usually quite sparse and we have to rely mainly on reports from others and particularly from television, newspapers and other media.

Combs & Slovic (1979) have shown empirically that newspapers report violent and homicidal events disproportionately and this results in a corresponding bias in public perceptions of their frequency. On the other hand, a most significant indirect input in these days is information from those scientists that are making what we have called objective assessments of risk. These are readily assimilated because they come from what psychologists call 'high credibility sources' and will increasingly narrow the subjective – objective gap to which I referred earlier.

There are a number of other factors that are thought to influence the perceptions of ordinary people; the distinction between a threat to the individual and one to society as a whole; the influence of voluntary acceptance of risk; the influence of familiarity and the factor of catastrophe size. The work of Green & Brown (1978) at Dundee and Slovic *et al.* (1980) in the United States is beginning to provide valuable empirical evidence, some of which will be discussed in more detail later in the programme. My own role will be to accept that cases have been made for the influence of these variables and to speculate on some of the underlying psychological mechanisms.

But first I should try to summarize briefly the methods used in the study of risk perception, for those who are not familiar with them.

Research methods in risk perception

The basic research method employed in risk perception requires the presentation of a list of hazards to samples of experimental subjects, in questionnaire form. The subject's task is to scale the hazards in order and magnitude of severity or to evaluate each hazard on a set of dimensions that are supplied.

In the simplest form of analysis, mean scores for each hazard are computed to derive a kind of average scale for the samples. This can then be compared directly with objectively derived scales. In more complex forms, what we may call the *patterns* of evaluation of the set of hazards are drawn out in the form of clusters of similar evaluation, principal components or dimensions. Recent use has also been made of multidimensional scaling analysis, which produces a two-dimensional or three-dimensional spatial mapping of relationships.

These approaches make it possible to compare different sets of hazards, different sets of people or, of course, evaluations of a single or set of hazards on a range of dimensions.

With a colleague, Kathy Rees, I have recently carried out an analysis of the perception of two samples of people towards sporting activities. The original aim was to determine how sports can be distinguished from non-sports. This is obviously a multidimensional perceptual problem and risk is one of the important dimensions that we measured.

Figure 1 shows a single computer resolution that gives the distribution in two-dimensional space of 38 activities, so that those activities with similar attributes are placed near to each other. We can observe those activities that have varying degrees of *risk*; we can see (by comparing the various partitionings for their congruence with that of risk) which other dimensions have high positive or negative correlations with riskiness and those that are independent of this attribute. The attributes included in this analysis in addition to *risk* are: skill, sense of achievement, relaxation, competitiveness, physical exertion, team involvement and degree of organization.

Alternatively, we can look at any single activity and see not only how risky it is, but how this quality is, comparatively speaking, combined with other qualities to form its unique meaning space.

The study was based on the responses of two samples, an 'expert' sample composed of members and staff of the Sports Council and a sample of the general

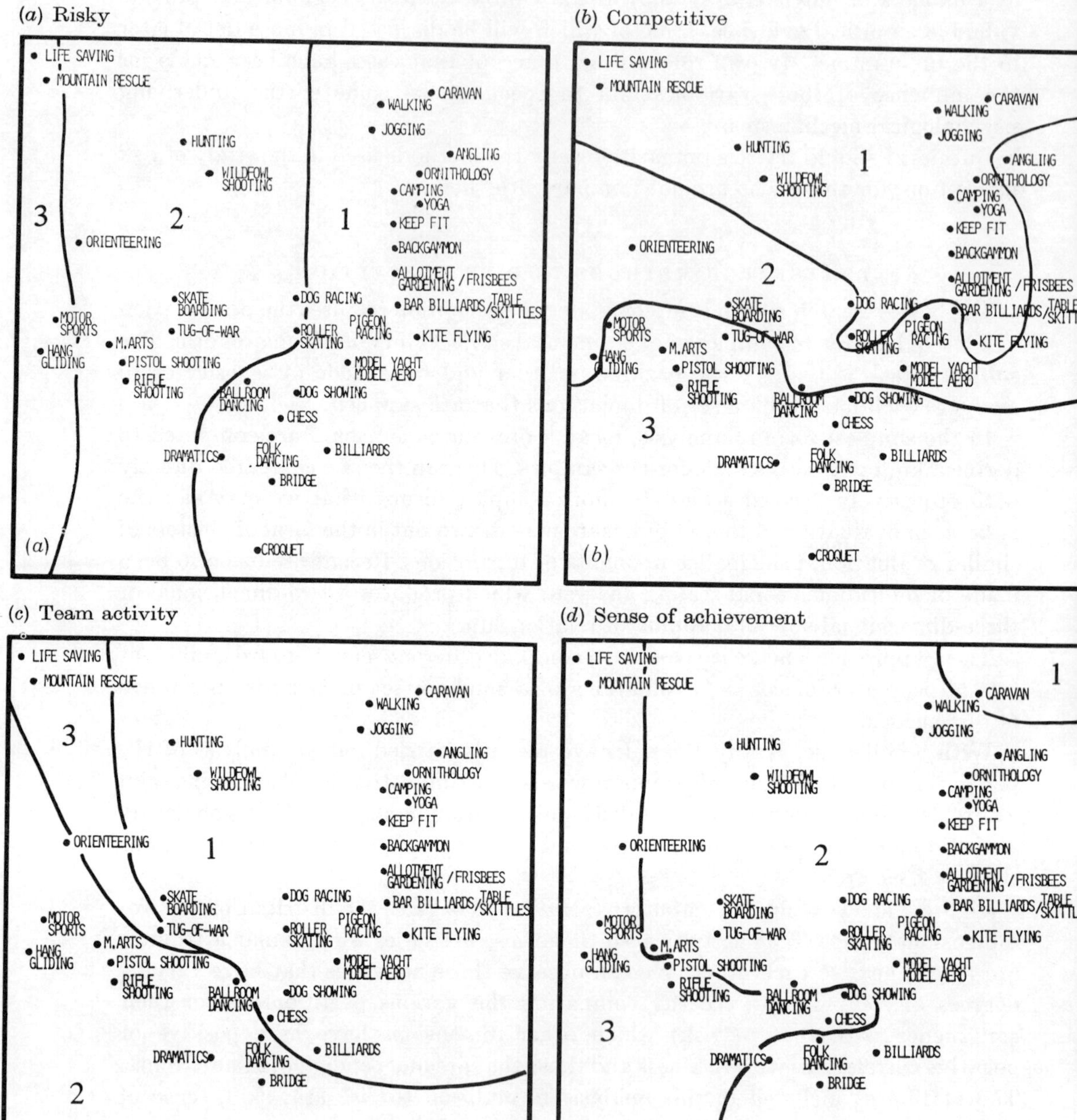

FIGURE 1. Multidimensional scalogram plots for eight attributes of 38 activities. Scale: 1, hardly at all; 2, moderately; 3, very. (*a*) Risky; (*b*) competitive; (*c*) team activity; (*d*) sense of achievement.

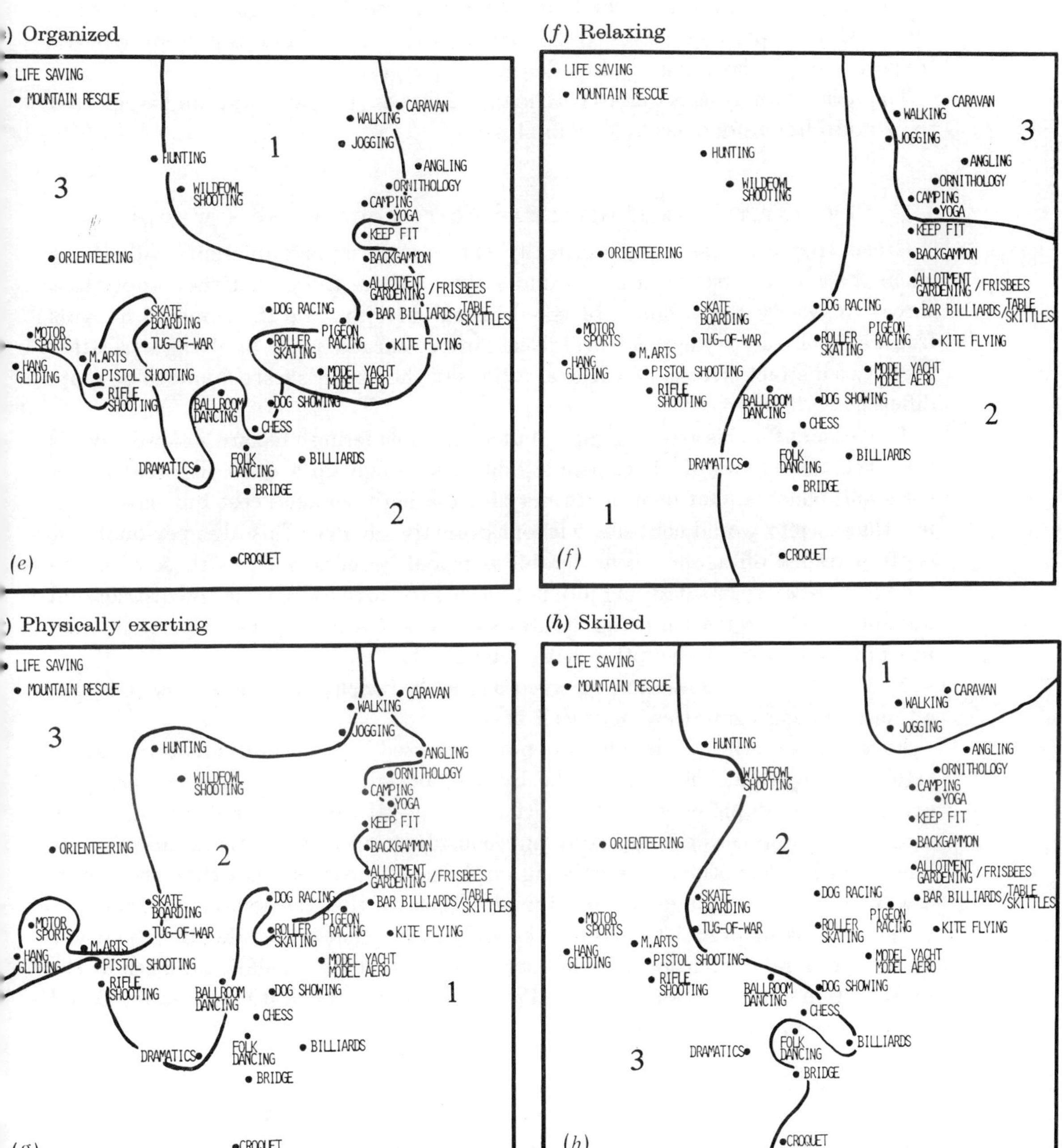

FIGURE 1 (*cont.*). (*e*) Organized; (*f*) relaxing; (*g*) physically exerting; (*h*) skilled.

public. The comparison between two such samples could usefully be generalized to other risk-perception contexts. Regrettably there is space only to present here the 'expert' sample data, mainly for illustrative purposes.

The form of analysis is due to Guttman & Lingoes (Lingoes 1973) and is referred to as multidimensional scalogram analysis.

Threat to the individual or to society as a whole

Green (1979) has shown empirically that people apply different evaluations when their own personal safety is under threat, from those that they apply to a threat to society as a whole. He asked his subjects to scale the same 15 hazards from snakebite up to the accidental release of nuclear radiation, applying separately these two distinct criteria. Their evaluations of the hazards were found to be quite different.

The reasons for this are not simple, but when considering a hazard's consequences from the two points of view, people doubtless weigh up a quite different set of costs and benefits. Our own death may have a high personal cost but most of us feel that society would continue without dramatic change. This high personal cost is often traded off against considerable personal benefits, such as those of maintaining a lucrative or satisfying job, not having to move away from the district and disrupt the children's schooling and, in some cases, a certain social prestige from pursuing a dangerous occupation with courage and 'cool'.

None of these benefits are perceived as vastly beneficial to society as a whole, although in aggregate they may well be so.

Furthermore, the trade-offs are not processed by an individual in simple algebraical ways as they have to be by cost–benefit analysts. There is plenty of evidence that people select and distort what they attend to, to reduce conflict, in some cases suppressing completely any consideration of the unpalatable aspects of a hazard and in others reconstruing unpleasant aspects so that they are seen in a way that is more congenial to the main thrust of their personal evaluation.

When thinking of society it is not the self concept that is at stake, but the whole system of moral evaluation that we have been taught since childhood, such as the preservation of civilization, the sanctity of human life, the destiny of mankind and our duty to unborn generations.

Voluntary and involuntary risk

It is often claimed that people perceive a risk as less serious if they accept it voluntarily. This is related to the distinction between a threat to personal safety and one to the wider society, because voluntary risks more often involve personal safety. The contraceptive pill is a good example. In fact, it is a very good example because acceptance of the personal risk may actually minimize the dangers facing society from overpopulation.

However, the correlation between voluntariness and personal safety is only moderate, for there are plenty of involuntary risks that involve threats to personal safety, such as conscription into the armed services.

Voluntariness has to do with whether you blame yourself or others and, if the latter, whom, when a calamity occurs. As such, it can be considered within a wider framework developed by psychologists and called attribution theory. This began with Fritz Heider's (1958) observation that 'it makes a real difference, whether a person discovers that the stick that strikes him falls from a rotting tree or was hurled by an enemy'.

If the cause is human and the event is an adverse one, we tend to attribute responsibility, and hence blame, proportionally to the various agents. In almost all cases, we have to reserve some portion of responsibility to ourselves for, as it were, 'letting them do it in the first place'. In some cases we are forced to acknowledge that the event was almost wholly our own fault. The point to emphasize is that the perceptual process of construction, the 'effort after meaning' is here being applied not merely to the event itself but to the whole series of causal antecedents. This happens in the most trivial of everyday happenings, but it is seen in its most ritualized form in the probing public enquiries that follow major catastrophes.

Jones & Davis (1961) claim that we infer from an event whether the agent had knowledge of what he was doing and whether he had the ability to avoid doing it; then before this, whether he had a deliberate intention to do it (which, of course, makes the event seem much more serious); and, finally, whether this intention sprang from a consistent intention to do this kind of thing.

If we are already ill-disposed towards the agent because we perceive his authority to control events to have been unilaterally imposed upon us (as some people perceive governments, scientists and captains of industry), much stronger blame is attributed. Another factor is the extent to which the agent has drawn some advantage (which we did not share) from the risk that he took on our behalf.

Two extensions of this attribution process are worth mentioning briefly. The first has been labelled 'locus of control' by Rotter (1966). People differ in the extent to which they tend consistently to attribute events to their own behaviour or see them as a function of external forces beyond their control.

The second phenomenon is 'learned helplessness'. Generalizing from the behaviour of experimental dogs that had been given a series of shocks from which they could not escape, Seligman & Maier (1967) have coined this term to describe the attitudes of impotence and feelings of depression that tend to nullify sensible coping behaviour under risk (Seligman 1975).

Attribution affects not only risk perception but risk management. There is evidence from experimental work, if it were needed, to confirm that when an outside agent allows or encourages us to become involved in the decision-making or control process, we like him better and are less likely to attribute blame if things should go wrong.

The implications for public consultation and participation are huge, but those responsible for major risks understandably become totally preoccupied with the technical means of reducing the risk and do not spend much time on the psychological means of rendering it realistically acceptable.

There is a preference for not rocking the boat, not lifting the lid, in the almost certainly mistaken belief that protest will be minimized.

Familiarity with the hazard

Another factor that is often thought to influence public perception is people's degree of familiarity with a hazard. The earlier research on natural hazards implied that farmers who had become accustomed to living on a flood plain greatly underestimated the threat (White 1952). The high levels of risk in smoking and in crossing the road are often thought to be sustained because they are so commonplace.

Recent research by Slovic *et al.* (1980) casts some doubt on the notion of a simple linear relation and, though it may appear unhelpful, the truth is probably that some risks are perceived to be worse and others better with increased familiarity. Their factor analysis of the perceptions of 90 hazards in terms of 18 risk characteristics revealed a factor of 'familiarity, observability, knowledge, etc.' that was relatively independent of the main factor, which encompassed such characteristics as 'dread, fatal, catastrophic, etc.'

Again, there are multiple dimensions involved and it is only by a more analytic approach that we shall make progress.

We have an innate fear response which, before it is overlaid by learning, is evoked by innate stimuli such as pain, loss of support and very loud noises. With repeated stimulation, the fear response subsides. This process is called habituation.

In the type of learning called conditioning, made famous by Pavlov's salivating dogs, in which a previously neutral stimulus, such as a green light, is paired with, say, pain, to produce a fear response, the new learning that takes place will be gradually extinguished if the originally provoking stimulus (that is, pain) is removed. This extinction or partial extinction often occurs because the animal learns a coping action: behaviour to avoid or escape the pain as soon as he sees the green light.

Finally, if a person learns a behaviour sequence that successfully operates on the environment to avoid or to remove the fear-provoking stimulus, the behaviour will be sustained and the fear and its anticipation removed.

Unfortunately, as we know from psychopathology, escape and avoidance behaviour can be maladaptive or even phobic. It may also lead to secondary gains, as where the person who initially takes the coping action of joining a movement to campaign against a feared hazard finds a new social warmth, a fellowship and social prestige in direct proportion to the intensity of his campaigning. So he intensifies his campaigning.

Guedeney & Mendel (1973), reporting on a local attitude survey of a nuclear power station in France, showed that anxiety is lower among those living near the plant. They refer to 'mastery through vision' by these residents, who are 'constantly observing everyday life and gradually adapting themselves to unconscious fears'.

In such cases, as in many others, there is repeated exposure to the learned fear stimulus but the catastrophe does not happen so the power of the stimulus to evoke the fear response is gradually weakened.

In other cases, as for example in the physician's or surgeon's increased familiarization with patients who are heavy smokers, the reverse occurs.

There are other hazards, of which high-risk occupations are a good example, where increased familiarity with the situation provides enhanced reinforcement from the *benefits* as well as the costs, so that the trade-off moves further into credit. Also, in the course of perception, as mentioned earlier, the risk to health may be reconstrued to resolve the conflict or the cognitive dissonance that it creates with, for example, the high pay and good working conditions.

Catastrophe size

The public is much affected in its perception of the seriousness of a hazard by the potential size of a single catastrophe. There is little doubt that this is one of the most important factors to account for the public's assessment of risk from nuclear power generation. They accept that an accident is highly unlikely but they perceive, rightly or wrongly, that it might be devastatingly serious if it occurred.

The critical factor appears to be the *simultaneous* loss of life, although in some hazards the total number of deaths may be very small and one could even suggest that if these deaths are inevitable it is tidier to have them in one package.

The explanation probably lies partly in the tendency of perception to simplify experience by selecting distinctive prototypes to signify what would otherwise be a motley aggregation of experiences. For example, our awareness that coal mining is a dangerous occupation is much more simply represented by images of the large-scale mining disaster than by cases of chronic pneumoconiosis.

Another factor is that a large-scale catastrophe is a vivid reminder that our continuing individual and societal efforts to control our environment, to cope with the brooding forces of Nature, is far from successful. The deep mythology of a world that has been created by, and can therefore be ended by, an all-powerful deity or overwhelmed by man's own greed or ineptitude is evoked. This is not irrational; it is perception at a somewhat more abstract level than the normal run of risk assessment.

Conclusion

In conclusion, then, we may ask if it is fair to say that the public's perceptions are irrational?

Certainly ordinary people rely on partial information, imperfect memories and distorted time perspectives to extrapolate from past experiences into the future.

This means that their probability predictions are doubtless inaccurate. But they do show rationality in their support of scientists who have assumed the task of improving such estimates and they seem willing to assimilate this new information.

Furthermore, they would seem to be the best judges of which variables to enter into the assessment equations and how they should be weighted. They apply reasoning to the trade-off of benefits against costs in arriving at their judgements. They include variables that are almost beyond quantification, but to which they are sensitized by their anxieties; they can extend this consideration to the cultural myths and moral values that they wish to preserve as members of a society.

We have to conclude, I think, that the public's perception of risk and so-called objective risk assessment are different but complementary forms of rationality and we should work towards their synchronization.

References (Lee)

Bartlett, F. C. 1932 *Remembering*. Cambridge University Press.

Combs, B. & Slovic, P. 1979 *Journalism Q.* **56**, 837–843.

Green, C. H. & Brown, R. A. 1978 *J. occup. Accid.* **2**, 55–70.

Green, C. H. 1979 Presented at Public Perception of Risk (Symposium on the Acceptability of Risk), U.M.I.S.T./University of Manchester, December.

Guedeney, C. & Mendel, G. 1973 *L'angoisse atomique et les centrales nucléaires*. Paris: Payot.

Heider, F. 1958 *The psychology of interpersonal relations*. New York: Wiley.

Jones, E. E. & Davis, K. E. 1961 *J. abnorm. soc. Psychol.* **63**, 302–310.

Lingoes, J. C. 1973 *The Guttman–Lingoes non-metric program series*. Ann Arbor, Michigan: Mathesis.

Rotter, J. B. 1966 *Psychol. Monogr.* no. 80.

Seligman, M. E. P. 1975 *Helplessness*. San Francisco: Freeman.

Seligman, M. E. P. & Maier, S. F. 1967 *J. exp. Psychol.* **74**, 1–9.

Slovic, P., Fischhoff, B. & Lichtenstein, S. 1979 *Environment* **21**, 14–20; 36–39.

Slovic, P., Fischhoff, B. & Lichtenstein, S. 1980 In *Societal risk assessment: how safe is safe enough?* (ed. R. Schwing & W. A. Albers), pp. 181–214. New York: Plenum.

Starr, C. 1969 *Science, N.Y.* **165**, 1232–1238.

White, G. F. 1952 *Human adjustment to floods: a geographical approach to the flood problem in the United States*. Research Paper no. 29, Department of Geography, University of Chicago.

Proc. R. Soc. Lond. A **376**, 17–34 (1981)
Printed in Great Britain

Perceived risk: psychological factors and social implications

BY P. SLOVIC, B. FISCHHOFF AND SARAH LICHTENSTEIN
Decision Research, 1201 *Oak Street*, *Eugene*, *Oregon* 97401, *U.S.A.*

Subjective judgements, whether by experts or lay people, are a major component in any risk assessment. If such judgements are faulty, risk management efforts are likely to be misdirected. This paper begins with an analysis of biases exhibited by lay people and experts when they make judgements about risk. Next, the similarities and differences between lay and expert evaluations are examined in the context of a specific set of hazardous activities and technologies. Finally, insights from this research are applied to the problems of informing people about risk and forecasting public response towards nuclear power.

People respond to the hazards that they perceive. If their perceptions are faulty, efforts at personal, public and environmental protection are likely to be misdirected. For some hazards, such as motor vehicle accidents, extensive statistical data are available to guide perceptions. For other familiar activities, such as the consumption of alcohol and tobacco, assessment of risk requires complex epidemiological and experimental studies. Still other hazards, such as those associated with nuclear power, are so new that risk assessment must be based on complex theoretical analyses such as fault trees, rather than on direct experience.

Despite an appearance of objectivity, all forms of risk assessment include a large component of subjective judgement. Someone, relying on educated intuition, must determine the structure of the problem, decide the consequences to be considered, and select the relevant data and interpret it. Once the analyses have been performed, they must be communicated to those who actually manage hazards, including industrialists, environmentalists, regulators, legislators and voters. If these people do not understand or believe the data they are shown, then distrust, conflict and ineffective hazard management are likely.

This paper explores some psychological elements of the risk assessment process. Its basic premises are that both the public and the experts are necessary participants in that process, that assessment is inherently subjective and prone to distortion due to judgemental limitations, and that understanding perceptions is crucial to effective decision making.

JUDGEMENTAL BIASES IN RISK PERCEPTION

When lay people are asked to evaluate risks, they seldom have statistical evidence on hand. In most cases, they must make inferences based on what they remember hearing or observing about the risk in question. Research has identified a number of general inferential rules that people use in such situations. These rules, known as *heuristics*, are employed to reduce difficult mental tasks to simpler ones. Although they are valid in some circumstances, in others they lead to large and persistent biases with serious implications.

Availability

One inferential strategy that has special relevance for risk perception is the *availability heuristic* (Tversky & Kahneman 1973). People using this heuristic judge an event as likely or frequent if instances of it are easy to imagine or recall. Because frequently occurring events are generally easier to imagine and recall than are rare events, availability is often an appropriate cue. However, availability is also affected by numerous factors unrelated to frequency of occurrence. For example, a recent disaster or a vivid film could seriously bias risk judgements.

Availability bias helps to explain people's misperceptions and faulty decisions with regard to certain natural hazards. Kates (1962) observed that residents of flood plains appeared to be 'prisoners of their experience', unable to conceptualize floods that have never occurred or to see the future as anything but a mirror of the recent past.

One particularly important implication of the availability heuristic is that discussing a low-probability hazard may increase its imaginability and hence its perceived riskiness, regardless of what the evidence indicates. For example, leaders in the field of recombinant DNA research quickly regretted ever bringing to public attention the remote risks of contamination by newly created organisms. Many discussions of the issue completely lost sight of the fact that the dangers were hypothetical and assumed that recombinant DNA laboratories were full of raging beasts. Ultimately, the very scientists whose concerns had initiated these discussions were vilified.

Judged frequency of lethal events

Availability bias is illustrated by several studies in which people judged the frequency of 41 causes of death (Lichtenstein *et al.* 1978). In one study, these people were first told the annual death toll in the United States (50000) for one cause (motor vehicle accidents) and then asked to estimate the frequency of the other 40. Figure 1 compares the judged number of deaths per year with the number reported in public health statistics. If the frequency judgements equalled the statistical rates, all data points would fall on the identity line. Although more likely hazards generally evoked higher estimates, the points were scattered about a curved line that lay sometimes above and sometimes below the line of accurate judgement. In general, rare causes of death were overestimated and common causes of death were underestimated. In addition to this general bias, sizable specific biases are evident in figure 1. For example, accidents were judged to cause as many deaths as diseases, whereas diseases actually take about 15 times as many lives. Homicides were incorrectly judged as more frequent than diabetes and stomach cancer deaths. Pregnancies, births and abortions were judged to take about as many lives as diabetes, though diabetes actually causes about 80 times more deaths. In keeping with availability considerations, overestimated causes of death (relative to the curved line) tended to be dramatic and sensational (accidents, natural disasters, fires, homicides), whereas underestimated causes tended to be unspectacular events, which claim one victim at a time and are common in non-fatal form (e.g. smallpox vaccinations, stroke, diabetes, emphysema).

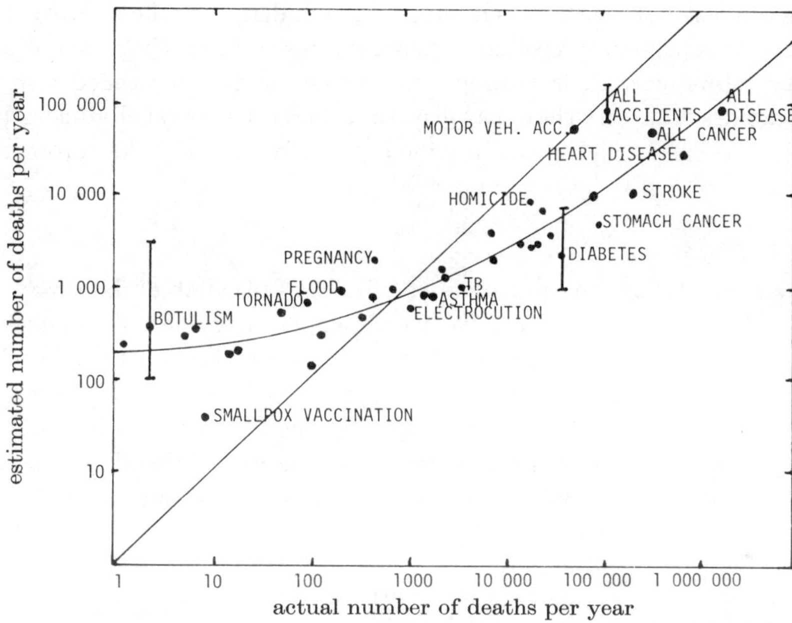

FIGURE 1. Relation between judged frequency and the actual number of deaths per year for 41 causes of death. Source: Lichtenstein *et al.* (1978).

Biased newspaper coverage and biased judgements

The availability heuristic highlights the vital role of experience as a determinant of perceived risk. If one's experiences are biased, one's perceptions are likely to be inaccurate. Unfortunately, much of the information to which people are exposed provides a distorted picture of the world of hazards. Following the study described above, Combs & Slovic (1979) examined the reporting of causes of death in two newspapers on opposite coasts of the United States. Both newspapers exhibited similar biases in their coverage of life-threatening events. For example, violent, often catastrophic, events were reported much more frequently than less dramatic causes of death with similar (or even greater) statistical frequencies. Moreover, these biases in newspaper coverage closely matched the biases in people's perceptions, discovered in our earlier studies.

It won't happen to me

Misleading experiences might also underlie another apparent judgemental bias, people's predilection to view themselves as personally immune to many hazards. Research shows that the great majority of individuals believe themselves to be better than average drivers, more likely than average to live past 80 years old, less likely than average to be harmed by products that they use, and so on. Although such perceptions are obviously unrealistic, the risks may look very small from the perspective of each individual's experience. Consider automobile driving: despite driving too fast, following too closely, etc., poor drivers make trip after trip without mishap. This personal experience demonstrates to them their exceptional skill and safety. Moreover, their indirect experience via the news media shows that when accidents happen, they happen to others. Given such misleading experiences, people may feel quite justified in refusing to take protective actions such as wearing seat belts (Slovic *et al.* 1978).

Out of sight, out of mind

In some situations, failure to appreciate the limits of 'available' data may lull people into complacency. For example, we asked people to evaluate the completeness of a fault tree showing the problems that could cause a car not to start when the ignition key was turned (Fischhoff *et al.* 1978). Respondents' judgements of completeness were about the same when looking at the full tree as when looking at a tree in which half of the causes of starting failure were deleted. In keeping with the availability heuristic, what was out of sight was also out of mind.

Overconfidence

Knowing with certainty

A particularly pernicious aspect of heuristics is that people typically have too much confidence in judgements based upon them. In another follow-up to the study on causes of death, people were asked to indicate the odds that they were correct in choosing the more frequent of two lethal events (Fischhoff *et al.* 1977). Odds of 100:1 or greater were given often (25 % of the time). However, about one out of every eight answers associated with such extreme confidence was wrong (fewer than 1 in 100 would have been wrong had the odds been appropriate). At odds of 10000:1 people were wrong about 10 % of the time. The psychological basis for this unwarranted certainty seems to be an insensitivity to the tenuousness of the assumptions upon which one's judgements are based. For example, extreme confidence in the incorrect assertion that homicides are more frequent than suicides may occur because people fail to appreciate that the greater ease of recalling instances of homicides is an imperfect basis for inference.

Hyperprecision

Overconfidence manifests itself in other ways as well. A typical task in estimating uncertain quantities such as failure rates is to set upper and lower bounds so that there is a certain fixed probability that the true value lies between them. Experiments with diverse groups of people making many different kinds of judgement have found that true values tend to lie outside of the confidence boundaries much too often. Results with 98 % bounds are typical. Rather than 2 % of the true values falling outside such bounds, 20–50 % usually do so (Lichtenstein *et al.* 1977). Thus people think that they can estimate uncertain quantities with much greater precision than they actually can.

Overconfident experts

Unfortunately, once they are forced to go beyond their data and rely on judgement, experts may be as prone to overconfidence as lay people. Fischhoff *et al.* (1978) repeated their fault-tree study with professional automobile mechanics (averaging about 15 years of experience) and found them to be about as insensitive as lay persons to deletions from the tree. Hynes & Vanmarcke (1976) asked seven 'internationally known' geotechnical engineers to predict the height of an embankment that would cause a clay foundation to fail and to specify confidence bounds around this estimate that were wide enough to have a 50 % chance of enclosing the true failure height. None of the bounds specified by these individuals actually enclosed the true failure height.

Further evidence of expert overconfidence may be found in many technical risk assessments. For example, an official review of the Reactor Safety Study concluded that despite the study's careful attempt to calculate the probability of a core meltdown in a nuclear reactor, 'we are certain that the error bands are understated. We cannot say by how much. Reasons for this include an inadequate data base, a poor statistical treatment [and] an inconsistent propagation of uncertainties throughout the calculation' (U.S. Nuclear Regulatory Commission 1978, p. vi). The 1976 collapse of the Teton Dam provides another case in point. The Committee on Government Operations attributed this disaster to the unwarranted confidence of engineers who were absolutely certain that they had solved the many serious problems that arose during construction (U.S. Government 1976).

Reconciling divergent opinions about risk

Both casual observation of risk debates and systematic empirical data suggest that experts and lay people have quite different perceptions about the riskiness of various technologies. Given that they are prisoners of rather different experiences, such divergence is to be expected. One would like to believe that, as evidence accumulates and the public and the experts come to share a common experience, their perceptions would converge towards one 'appropriate' view. Unfortunately, this is not likely to be so. A great deal of research indicates that, once formed, people's beliefs change very slowly, and are extraordinarily persistent in the face of contrary evidence (Nisbett & Ross 1980). Initial impressions tend to structure the way that subsequent evidence is interpreted. New evidence appears reliable and informative if it is consistent with one's initial beliefs, whereas contrary evidence is dismissed as unreliable, erroneous or unrepresentative.

Characterizing perceived risk

If it is to aid hazard management, a theory of perceived risk must explain people's extreme aversion to some hazards, their indifference to others, and the discrepancies between these reactions and experts' recommendations. Why, for example, do some communities react vigorously against the location of a liquid natural gas terminal in their vicinity, despite the assurances of experts that it is safe? Why, on the other hand, do many communities situated on earthquake faults or below great dams show little concern for experts' warnings? Over the past few years we have been attempting to answer such questions as these by examining the opinions that people express when they are asked, in a variety of ways, to characterize and evaluate hazardous activities and technologies (Slovic *et al.* 1979*a*, *b*, 1980*a*). This descriptive research aims (*a*) to develop a taxonomy of risk characteristics that can be used to understand and predict societal responses to hazards, and (*b*) to develop methods for assessing public opinions about risk in a way that could be useful for policy decisions.

Quantifying perceived risk

In one study, we asked four different groups of people to judge 30 hazardous activities, substances and technologies according to the 'present risk of death from each (across U.S. society as a whole)'. Three groups were from Eugene, Oregon; they included 69 college students, 76 members of the League of Women Voters (LOWV), and 47 business and professional members of the 'Active Club'. The fourth group was composed of 15 experts in risk assessment.

Table 1 rank orders the mean risk judgements for the four groups. There were many similarities between the three groups of lay people. For example, each group believed that motorcycles, motor vehicles and handguns were highly risky, while vaccinations, home appliances, power mowers, and football posed relatively little risk. However, there were strong differences as well. Active Club members viewed pesticides as much less risky than did the other groups. Nuclear power was rated as highest in risk by the League of Women Voters and student groups, but only eighth by the Active Club. The students viewed contraceptives as riskier and mountain climbing as safer than did the other lay groups. Experts' judgements of risk differed markedly from the judgements of lay people. The experts viewed electric power, surgery, swimming and X-rays as more risky than did the other groups and they judged nuclear power, police work and mountain climbing to be much less risky.

What determines risk perception

What did people mean, in this study, when they said that a particular technology was quite risky? A series of additional studies was addressed to this question.

TABLE 1. ORDERING OF PERCEIVED RISK FOR 30 ACTIVITIES AND TECHNOLOGIES

(The ordering is based on the geometric mean risk ratings within each group. Rank 1 represents the most risky activity or technology.)

	League of Women Voters	college students	Active Club members	experts
nuclear power	1	1	8	20
motor vehicles	2	5	3	1
handguns	3	2	1	4
smoking	4	3	4	2
motorcycles	5	6	2	6
alcoholic beverages	6	7	5	3
general (private) aviation	7	15	11	12
police work	8	8	7	17
pesticides	9	4	15	8
surgery	10	11	9	5
fire fighting	11	10	6	18
large construction	12	14	13	13
hunting	13	18	10	23
spray cans	14	13	23	26
mountain climbing	15	22	12	29
bicycles	16	24	14	15
commercial aviation	17	16	18	16
electric power (non-nuclear)	18	19	19	9
swimming	19	30	17	10
contraceptives	20	9	22	11
skiing	21	25	16	30
X-rays	22	17	24	7
high school and college football	23	26	21	27
railroads	24	23	20	19
food preservatives	25	12	28	14
food colouring	26	20	30	21
power mowers	27	28	25	28
prescription antibiotics	28	21	26	24
home appliances	29	27	27	22
vaccinations	30	29	29	25

Perceived risk compared with frequency of death

When people judge risk, are they simply estimating frequency of death? To answer this question, we compared the risk judgements with technical estimates of the annual number of deaths for these hazards. The experts' judgements of risk were so closely related to the statistical (or calculated) frequencies that it seemed reasonable to conclude that they had good knowledge of the technical estimates and viewed 'risk' as synonymous with these estimates. The risk judgements of lay people were, however, only moderately related to the annual death rates, raising the possibility that risk means something different to them.

Lay fatality estimates

Perhaps lay people base their risk judgements on subjective fatality estimates that are inaccurate. To test this hypothesis, we asked additional groups of students and League of Women Voters members to estimate how many people were likely to die in the U.S. in the next year (if the next year was an average year) as a consequence of these 30 hazardous activities. If lay people equate risk with annual fatalities, their subjective fatality estimates, no matter how inaccurate, should resemble their risk judgements. We found, however, only a modest agreement between these two sets of judgements. Lay people's risk perceptions were, in fact, no more closely related to their own fatality estimates than they were to the technical estimates. Apparently, lay people incorporate other considerations besides annual fatalities into their concept of risk.

Disaster potential

One clue to these other considerations was the fact that the League of Women Voters members and students judged nuclear power as higher in risk and lowest in annual fatalities. One possible explanation is that they considered nuclear power to be a high-risk technology because of its potential for disaster. We therefore asked these same respondents to indicate, for each hazard, 'how many times more deaths would occur if next year were particularly disastrous rather than average'. For most hazards these multipliers were quite small, indicating that people saw little potential for disaster. The striking exception was nuclear power. Applying each person's disaster multiplier to their estimated fatalities for an average year, we found that almost 40 % of the respondents expected more than 10000 fatalities from nuclear power if next year were a disastrous year. More than 25 % expected 100000 or more fatalities.

Disaster potential seems to explain the discrepancy between the perceived risk and the annual fatality estimates for nuclear power. Yet, because disaster was judged significant for only a few of the hazards, it provided only a partial explanation of the perceived-risk data.

Risk characteristics

Our search for additional knowledge about risk perception led us to ask experts, students, League of Women Voters members and Active Club members to rate the 30 items on nine characteristics that had been hypothesized as relevant to individual and societal reactions to hazards. These characteristics included the degree to which an activity's risks were voluntary, controllable, known to science, known to those exposed, familiar, dread, certain to be fatal, catastrophic, and immediately manifested.

Mean ratings were quite similar for all four groups. Particularly interesting was the characterization of nuclear power, which had the dubious distinction of scoring at or near the extreme on all of the undesirable characteristics. Its risks were seen as involuntary, delayed, unknown, uncontrollable, unfamiliar, catastrophic, dread and fatal. This contrasted sharply with the characterizations of non-nuclear electric power and another radiation technology, X-rays. Electric power and X-rays were both judged more voluntary, less certain to be fatal, less catastrophic, less dreaded, more familiar, and less risky than nuclear power.

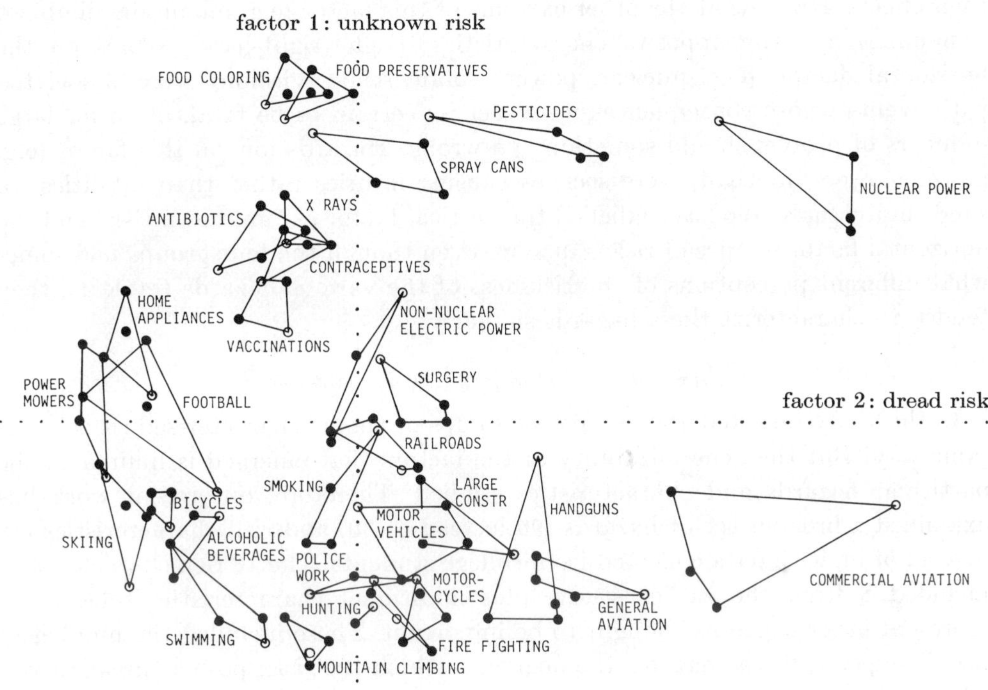

FIGURE 2. Location of 30 hazards within the two-factor space obtained from League of Women Voters, student, Active Club and expert groups. Connected lines join or enclose the loci of four group points for each hazard. Open circles represent data from the expert group. Unattached points represent groups that fall within the triangle created by the other three groups.

Across all 30 hazards, ratings of dread and of the likelihood of a mishap being fatal were closely related to lay judgements of risk. In fact, the risk judgements of the League of Women Voters and student groups could be predicted almost perfectly from ratings of dread and lethality and the subjective fatality estimates for normal and disastrous years. Experts' judgements of risk were not related to any of the nine risk characteristics.

Many pairs of risk characteristics tended to be correlated with each other across the 30 activities and technologies. For example, risks faced voluntarily were typically judged well known and controllable. These interrelations were sufficiently high to suggest that all the ratings could be explained in terms of a few basic dimensions of risk. To identify such dimensions, we conducted a factor analysis of the correlations from each group (principal components analysis with varimax rotation to simple structure). We found that the nine characteristics could be represented by two underlying factors that appeared to be the same for each group. Figure 2 illustrates the factor scores for each hazard within the common space. Hazards at the high end of the vertical dimension or factor (e.g. food colouring, pesticides) tended to be new, unknown, involuntary, and delayed in their effects. Hazards at the other extreme of this factor (e.g. mountain climbing, swimming) had the opposite characteristics. High (right-hand) scores on the horizontal factor (e.g. nuclear power, commercial aviation) were associated with events whose consequences were seen as certain to be fatal, often for large numbers of people, should something go wrong. Hazards low on this factor (e.g. power mowers, football) were seen as causing injuries rather than fatalities, to single individuals. We have labelled the vertical factor as 'unknown risk' and the horizontal factor as 'dread risk'. In sum, even though the four groups had somewhat different perceptions of the riskiness of the various hazards (table 1), they tended to characterize these hazards similarly.

An extended study of risk perception

In the study reported above, diverse groups of people characterized risks in the same way. But the generalizability of the factors that emerged is limited to the particular hazards and characteristics studied. Therefore, our recent work has examined a broader set of hazards (90 instead of 30) and risk characteristics (18 instead of 9), with data collected from college students. The 18 risk characteristics included 8 from the earlier study, plus additional characteristics selected to represent other concerns thought to be important. These included the number of people exposed to the hazard, the degree to which the risks pose a threat to you (the rater) personally, threat to future generations, and threat of global catastrophe. We also asked about the degree to which the activity's benefits are equitably distributed to those who bear the risks, the observability of the damage-producing processes, the degree to which the risks are increasing and the ease of reducing the risks.

Each of the 90 hazards was rated on overall riskiness and judged on all 18 characteristics of risk. In general, the risks from most of these activities were judged to be increasing, not easily reduced, and better known to science than to those people exposed to them.

As in the earlier studies, many pairs of risk characteristics were highly correlated with each other. Factor analysis showed that the 18 characteristics could be represented well by three factors, the first two of which resembled the two factors that emerged from the earlier studies. Factor 1 was associated with lack of control, fatal consequences, high catastrophic potential, reactions of dread, inequitable distribution of risks and benefits (including transfer of risks to future generations), and the belief that the risks are increasing and not easily reducable. Factor 1 thus seems to correspond closely to the factor labelled 'dread risk' in the earlier study. Factor 2 was associated with risks that are unknown, unobservable, new, and delayed in their manifestation. It thus corresponded closely to factor 1 ('unknown risk') of the earlier study. Factor 3 was primarily determined by the number of people exposed and the rater's personal exposure.

Scores for the individual items on factors 1 and 2 are plotted in figure 3. The hazards at the extremes on each dimension give support to the factor names, which were initially determined from examination of the set of characteristics defining each factor. Items at the high end of factor 1 (nerve gas, nuclear power

FIGURE 3. Hazard locations on factors 1 and 2 of the three-dimensional structure derived from the interrelations among 18 risk characteristics. Factor 3 (not shown) reflects the number of people exposed to the hazard and the degree of one's personal exposure. Source: Slovic *et al.* (1980*a*).

accidents, nuclear weapons, terrorism, warfare and crime) are all highly dreaded, in contrast to the items at the opposite end (home appliances, bicycles, Christmas tree lights, hair dyes). The locations of items on the vertical axis correspond to the degree to which their risks are perceived as known, familiar and observable. Hazards falling at the high exposure end of factor 3 (societal and personal exposure) were motor vehicle accidents, caffeine, alcoholic beverages, smoking, food preservatives, herbicides and pesticides. Hazards falling at the low end on this factor included lasers, solar electricity, space exploration, laetrile, scuba diving and open-heart surgery.

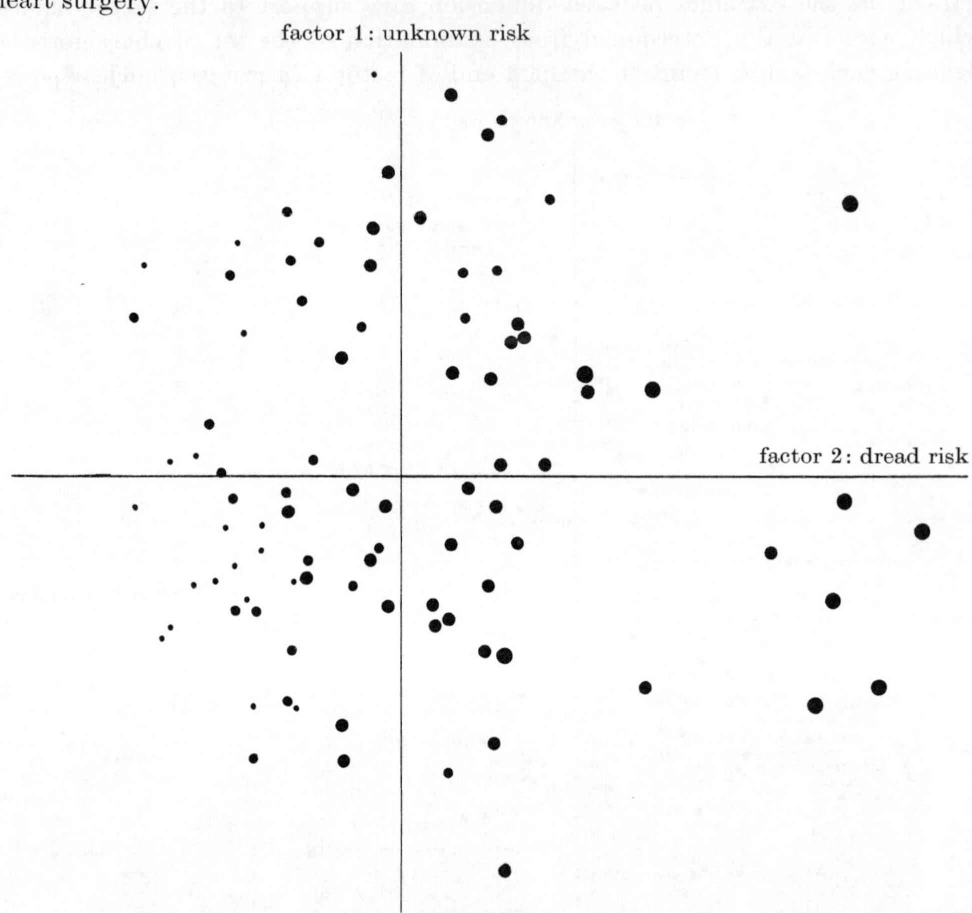

FIGURE 4. Attitudes towards regulation of the hazards shown in figure 3. The larger the dot, the greater the desire for strict regulation to reduce risk.

Our most recent study has examined 81 hazardous activities, including many more chemicals (e.g. 2,4,5-T, hexachlorophene, polyvinyl chloride) than were included in previous studies, and numerous hazards whose risks were disaggregated (e.g. motor vehicle risks were partitioned into four separate items treating (*a*) accidents to vehicle occupants, pedestrians and cyclists, (*b*) auto racing, (*c*) carbon monoxide exhaust, and (*d*) airborne lead). Factor analysis, based on ratings of these hazards on the 18 risk characteristics, resulted in three factors almost identical to those of the previous study.

We have found that lay people's risk perceptions and attitudes are closely related to the position of a hazard within the factor space. Most important is dread risk. The higher an activity's score on this factor, (*a*) the higher its perceived risk, (*b*) the more people want its risks reduced, and (*c*) the more they want to see strict regulation employed to achieve the desired reduction in risk (see figure 4). The attitudes and perceptions of experts, however, appear much less closely related to the factor space.

Social implications

In this section, we shall briefly describe two applied problems to which we believe this basic research to be relevant. One is the challenge of informing or educating people about risk; the second is the problem of understanding and forecasting public response towards new technologies, as illustrated by nuclear power. Our coverage here will be brief. Further details about these two applications can be found in Slovic *et al.* (1979*b*, 1980*b*).

Informing people about risk

One dramatic change in people's outlook on life in recent years is their growing awareness of the risks that they encounter in daily experience. The consequence of this awareness has been increased pressure on the designers and regulators of hazardous enterprises to inform people about the risks that they face. For example:

(1) the Food and Drug Administration is mandating patient information inserts for an increased number of prescription drugs;

(2) the Department of Housing and Urban Development now requires the sellers of homes built before 1950 to inform buyers about the presence of lead-based paints;

(3) the White House has directed the Secretary of Health, Education, and Welfare to develop a public information programme on the health effects of radiation exposure.

Despite good intentions, creating effective informational programmes may be quite difficult. Doing an adequate job means finding cogent ways of presenting complex technical material that is clouded by uncertainty and may be distorted by the listeners' preconceptions (and perhaps misconceptions) about the hazard and its consequences. For, as we have seen above, misleading personal experiences may sometimes promote a false sense of security, whereas in other circumstances, more discussion of possible adverse consequences may enhance their apparent threat. Moreover, research has also demonstrated that people's beliefs change slowly and show extraordinary persistence in the face of contrary evidence. What follows is a brief review of some additional problems that information programmes must confront.

Presentation format is important

Subtle changes in the way that risks are expressed can have a major impact on perceptions. For example, Slovic *et al.* (1978) argued that motorists' reluctance to wear seat belts might be due to the extremely small probability of incurring a fatal accident on a single automobile trip. Since a fatal accident occurs only about once in every 3.5×10^6 person-trips and a disabling injury only once in every 100000 person-trips, refusing to buckle one's seat belt may seem quite reasonable. It looks less reasonable, however, if one adopts a multiple-trip perspective and considers the substantial probability of an accident on some trip. Over 50 years of driving (about 40000 trips), the probability of being killed rises to 0.01 and the probability of experiencing at least one disabling injury is 0.33. We found that people asked to consider this lifetime perspective responded more favourably towards the use of seat belts (and air bags) than did people asked to consider a trip-by-trip perspective.

Numerous other format effects have been documented in the literature on risk assessment. For example, merely fusing or splitting branches in a fault tree (without adding or subtracting any information) affects people's perceptions; a given category of problems tends to be viewed as contributing more to failures when split into two branches than when presented as one branch. The same risk options, described in terms of lives saved, may be evaluated differently when framed in terms of lives lost. Details of these and other effects are given in Tversky & Kahneman (1981) and Slovic *et al.* (1980*b*).

The fact that subtle differences in how risks are presented can have marked effects on perceptions and actions suggests that those responsible for information programmes have considerable ability to manipulate perceptions. Moreover, since these effects are not widely known, people may inadvertently be manipulating their own perceptions by casual decisions that they make about how to organize their knowledge.

Cross-hazard comparisons may be misleading

One common approach to deepening people's perspectives is presenting quantified risk estimates for a variety of hazards. These presentations typically involve elaborate tables and even 'catalogues of risks' in which diverse indices of death or disability are displayed for a broad spectrum of life's hazards. Some of these provide extensive data on risks per hour of exposure, showing, for example, that an hour of riding a motorcycle is as risky as an hour of being 75 years old. One analyst developed lists of activities, each of which is estimated to increase one's chances of death (in any year) by 1 in 10^6. Other analysts have ranked hazards in terms of their expected reduction in life expectancy. Those who compile such data typically assume that they will be useful for decision making.

Research on perceived risk implies that comparisons such as these will not, by themselves, be adequate guides to personal or public decision policies. Risk perceptions and attitudes appear to be determined not only by accident probabilities, annual mortality rates or the mean losses of life expectancy, but also by numerous other characteristics of hazards such as uncertainty, controllability, catastrophic potential, equity and threat to future generations. Within the perceptual space defined by these and other characteristics, each hazard is unique. To many persons, statements such as 'the annual risk from living near a nuclear power plant is equivalent to the risk of riding an extra three miles in an automobile' appear ludicrous because they fail to give adequate consideration to the important differences in the nature of the risks from these two technologies.

Conclusions

The development of programmes to inform patients, workers and consumers about risk is an admirable goal. However, it is important to recognize the difficulties confronting such programmes. There is need for extensive empirical research on the problems of communicating information about risk. Since every decision about the content and format of an information statement is likely to influence perception and behaviour (and ultimately product viability, jobs, electricity costs, compliance with medical treatments, and other important consequences), extreme care must be taken to select knowledgeable and trustworthy designers and programme coordinators. Finally, it is important to recognize that informing people, whether by warning labels, package inserts or extensive media presentations, is but part of the larger problem of helping them to cope with the risks and uncertainties of modern life. We believe that much of the responsibility lies with the schools, whose curricula should include material designed to teach people that the world in which they live is probabilistic, not deterministic, and to help them learn judgement and decision strategies for dealing with that world. These strategies are as necessary for navigating in a world of uncertain information as geometry and trigonometry are to navigating among physical objects.

Forecasting public response: nuclear power

Research on risk perception can help hazard managers and policy makers to understand public attitudes and forecast future acceptance or rejection of new technologies. A case in point is nuclear power, whose isolated position in the perceptual space reflects our respondents' view that its risks are unknown, dread, uncontrollable, inequitable, catastrophic and likely to affect future generations. Once such concerns are identified, one can ask how likely they are to change over time in response to education, a good safety record, or an accident.

Basic perceptions

Studies of people opposed to nuclear power show that they judge its benefits as quite low and its risks as unacceptably great (Slovic *et al.* 1979*b*). On the benefit side, opponents do not see nuclear power as a vital link in meeting basic energy needs; rather, many view it as a supplement to other sources of energy, which are themselves adequate. On the risk side, nuclear power evokes extreme feelings of dread.

Though a number of commentators have speculated that people's strong fears of nuclear power stem from a belief that death from radiation is somehow more horrible than death from other causes, our studies indicate that their fears appear to derive instead from concern over how many deaths are likely. Many people's mental images of a nuclear accident include the spectre of hundreds of thousands, even millions, of immediate deaths, accompanied by incalculable and irreversible damage to the environment. These images bear little resemblance to the views of industry officials (and most technical experts), who expect redundant safety and containment systems to prevent almost all reactor accidents and limit the damage of those that do occur. Industry proponents have tended to attribute this perception gap to public ignorance and irrationality.

We question this attribution and we doubt that its proposed remedy, education, will easily succeed, for although people's fears may be exaggerated, they are not divorced from reality. On the technical side, the low probability of catastrophic nuclear mishaps makes demonstration of their improbability difficult from a statistical standpoint. Furthermore, people are aware that many of the 'facts' of nuclear risks are in dispute and that experts have been wrong in the past, as when they irradiated enlarged tonsils or permitted observers to witness atom-bomb tests at close range. On the psychological side, public concerns are triggered by reliance on memorability and imaginability (the availability heuristic). The risks of nuclear power would seem to be a prime candidate the availability bias because of the extensive media coverage that they receive and their association with the vivid, imaginable dangers of nuclear war.

One disturbing possibility is that objective discussions of nuclear safety may increase the imaginability of improbable mishaps, causing them, by the availability heuristic, to seem more likely. Consider an engineer arguing the safety of nuclear power by pointing out the improbability of the various ways radioactivity could be accidentally released: rather than reassuring the audience, the presentation might lead them to think, 'I didn't realize there were that many things that could go wrong.'

Whereas availability bias may contribute to the perception gap between pro-nuclear experts and their lay opponents, it does not point unambiguously to either side as having the most accurate appraisal of the risks from nuclear power. Although reliance on imaginability, by blurring the distinction between what is remotely possible and what is probable, is capable of enhancing public fears, inability to imagine all the possible ways that systems could fail might produce a false sense of security among technical experts. In so far as the actual risks may never be known with great precision and the interpretation of new information is strongly influenced by one's prior beliefs, the perception gap may be with us for a long time. Thus, the Three Mile Island accident 'proved' the possibility of a catastrophic meltdown to some, whereas to others it demonstrated the reliability of the multiple containment systems.

A nuclear future

Are the strong fears and determined opposition to nuclear power likely to persist? Will nuclear power ever gain widespread public acceptance? Although answers to these questions are by no means clear, public response to X-rays provides some clues. The almost universal acceptance of X-rays shows that a radiation technology can be tolerated if its use is familiar, its benefits clear, and its practitioners trusted. With nuclear power, the path to respectability exemplified by X-rays will have to be accompanied by an incontrovertible safety record. Research indicates that accidents occurring with unknown and potentially catastrophic technologies will be seen as signals that portend loss of control and indicate further, and possibly severe, losses (Slovic *et al.* 1980*a*). Thus even 'small' nuclear power accidents will probably have immense consequences for the industry and for society.

A quicker path to acceptance, and one that may provide the primary hope for the industry, could be forged by a severe energy shortage. Society has shown itself willing to accept increased risks in exchange for increased benefits. Brownouts, blackouts, or rationing of electricity would probably enhance the perceived need for nuclear power and increase public tolerance of its risks. Such crisis-induced acceptance of nuclear power may, however, produce anxiety, stress and conflict in a population forced to tolerate what it perceives as great risk because of its addiction to the benefits of electricity.

Conclusion

The study of human cognitive processes indicates that making intelligent decisions about risky activities is a very difficult task. It also raises a number of critical questions: Can society rise above the limitations of individual minds? Are new technologies forcing us to make decisions that we cannot make well (or successfully)? Should we take smaller steps in our technological development, so that we can recover from the inevitable mistakes? What kind of political institutions are needed to preserve democratic freedoms and ensure public participation for problems involving technical complexity, catastrophic risks and great uncertainty? If public debates and communications from experts do little to allay fears and, indeed, may exacerbate them, how should we structure public participation? What role can education play in helping society understand and cope with risk?

Although the study of individual human minds is a rather narrow starting point for examining societal risk decisions, it appears to lead quickly to important issues that need to be addressed by the entire community of scientists, policy makers and citizens.

This work was supported by the Technology, Assessment and Risk Analysis Program of the National Science Foundation under grant no. PRA79–11934 to Clark University under subcontract to Perceptronics, Inc. Any opinions, findings and conclusions expressed in this publication are those of the authors and do not necessarily reflect the views of the National Science Foundation.

References (Slovic *et al.*)

Combs, B. & Slovic, P. 1979 *Journalism Q.* **56**, 837–843; 849.

Fischhoff, B., Slovic, P. & Lichtenstein, S. 1977 *J. exp. Psychol.: hum. Percept. Perform.* **3**, 552–564.

Fischhoff, B., Slovic, P. & Lichtenstein, S. 1978 *J. exp. Psychol.: hum. Percept. Perform.* **4**, 330–344.

Hynes, M. & Vanmarcke, E. 1976 In *Proceedings of the ASCE Engineering Mechanics Division Speciality Conference*. Waterloo, Ontario: University of Waterloo Press.

Kates, R. W. 1962 *Hazard and choice perception in flood plain management*. University of Chicago, Department of Geography, Research Paper no. 78.

Lichtenstein, S., Fischhoff, B. & Phillips, L. D. 1977 In *Decision making and change in human affairs* (ed. H. Jungermann & G. deZeeuw), pp. 275–324. Amsterdam: D. Reidel.

Lichtenstein, S., Slovic, P., Fischhoff, B., Layman, M. & Combs, B. 1978 *J. exp. Psychol.: hum. Learn. Memory* **4**, 551–578.

Nisbett, R. & Ross, L. 1980 *Human inference*. Englewood Cliffs, N.J.: Prentice-Hall.

Slovic, P., Fischhoff, B. & Lichtenstein, S. 1978 *Accident Anal. Prev.* **10**, 281–285.

Slovic, P., Fischhoff, B. & Lichtenstein, S. 1979*a* *Environment* **21** (3), 14–20; 36–39.

Slovic, P., Fischhoff, B. & Lichtenstein, S. 1980*a* In *Societal risk assessment: How safe is safe enough?* (ed. R. Schwing & W. A. Albers, Jr), pp. 181–214. New York: Plenum.

Slovic, P., Fischhoff, B. & Lichtenstein, S. 1980*b* In *Product labeling and health risks* (ed. L. Morris, M. Mazis & I. Barofsky), pp. 165–181. Cold Spring Harbor, N.Y.: The Banbury Center.

Slovic, P., Lichtenstein, S. & Fischhoff, B. 1979*b* In *Energy risk management* (ed. G. T. Goodman & W. D. Rowe), pp. 223–245. London: Academic Press.

Tversky, A. & Kahneman, D. 1973 *Cognitive Psychol.* **4**, 207–232.

Tversky, A. & Kahneman, D. 1981 The framing of decisions and the rationality of choice. *Science, N.Y.* **211**, 453–458.

U.S. Government 1976 *Teton Dam disaster*. Washington, D.C.: Committee on Government Operations.

U.S. Nuclear Regulatory Commission 1978 *Risk assessment review group report to the U.S. Nuclear Regulatory Commission*, NUREG/CR-0400. Washington, D.C.: The Commission.

Discussion

F. J. C. Roe (19 *Marryat Road, London, U.K.*). In research on perception of risk, should not some attempt be made to ensure that, when lists are presented of factors to rank for risk, the lists used contain all the factors that the person questioned regards as important?

P. Slovic. Most studies to date have asked people to judge those characteristics that the researchers believed *a priori* might be important. Research is needed (and is being planned) in which the respondent gets an opportunity to indicate the characteristics that he or she believes to be relevant.

Proc. R. Soc. Lond. A **376**, 35–50 (1981)
Printed in Great Britain

Comparative risk perception: how the public perceives the risks and benefits of energy systems

By Kerry Thomas

Faculty of Social Science, The Open University, Walton Hall, Milton Keynes MK7 6AA, U.K.

The view of risk perception adopted in this paper focuses on how individuals define, and hence feel about, the outcomes of a risk issue. Perception of risk is seen as encompassing a variety of attributes of a risk issue including wider beliefs that permit a risk–benefit trade-off. The convergence between this differentiated view of risk perception and expectancy-value attitude theory was used to provide a technique for the description and comparison of risk perception across various risk issues, and between different, socially relevant, groups.

A survey of beliefs and attitudes of the general public toward the use of various energy systems (coal, oil, hydro, solar and nuclear energy) was carried out in Austria at a time of increasing concern with energy strategies and the controversy surrounding Austria's first nuclear energy plant.

The study showed that the public does conceive risk issues in differentiated terms, taking into account several substantive dimensions of both risk and probable benefits. While such dimensions might well be specific to the risk issue in question, it does seem likely that both risks and probable benefits will form part of belief systems in most instances where risk acceptance, or otherwise, is an issue.

In general, energy systems were found to be associated with only environmental risk; however, nuclear energy was an exception in that the public, whether in favour of nuclear energy (pro), or against it (con), believed that it is also associated with psychological and physical risk. Those against nuclear energy also believed that it is associated with indirect (future-oriented and political) risk, while those in favour did not associate nuclear energy with environmental risk. Overall, those members of the public with pro nuclear attitudes believed more strongly than those with con nuclear attitudes in the economic and technological benefits of energy generation in general.

The question was raised as to how an understanding of risk perception in the sense described here might be used in the service of policy making. It was suggested that, on issues that are the subject of public debate, policy makers might already possess sufficient information to understand the reasoning of public groups. This hypothesis was tested by asking a group of policy makers to take part in a role-play experiment. Their responses in the role of typical Austrian citizens either pro or con nuclear energy were compared with the actual responses of the public. In general, the policy makers were accurate, but there were some errors of attribution. In particular, the policy makers underestimated the public's negative feelings about the psychological risk dimension and their belief in its association with nuclear energy. This mistaken attribution was especially large for those members of the public who were in favour of nuclear energy.

1. Introduction

This paper explores the concept of risk perception and some of the implications of its essentially subjective viewpoint, drawing on evidence from a survey of beliefs and attitudes of the general public toward the use of various energy systems (coal, oil, hydro, solar and nuclear energy). The survey was carried out in Austria between 1977 and 1978 at a time of increasing concern with energy strategies and in particular the controversy surrounding Austria's first nuclear energy plant. The paper is organized around three themes.

(i) Risk perception as an idiosyncratic response in a complex world.

(ii) A convergence between risk perception and attitude theory, and the use of an attitude-based methodology for a comparative study of risk perception.

(iii) A discussion of some of the implications of adopting a subjective view of risk, in particular the accuracy of policy makers in their understanding of how different public groups perceive risk.

(a) *Risk perception*

Perception of risk, like all perception, is an idiosyncratic process of interpretation, a process of making sense of a complex world in order to plan, choose and act in that world. Risk perception is therefore assumed to involve a subjective probability judgement about the occurrence of an unpleasant event. Kahneman & Tversky, among others, have described some of the systematic biases that occur when individuals make judgements under uncertainty (Kahneman & Tversky 1973; Tversky & Kahneman 1974). But more relevant to this paper is Tversky's argument that the whole notion of rational choice under risk (including axiomatic utility theory when used as a descriptive model of an individual's choice behaviour) depends critically on the individual's interpretation of the consequences (Tversky 1974). Thus the way an individual perceives a risk will depend on just how *he* defines, and hence feels about, the outcome. And this definition may be considerably more differentiated than that envisaged by the experts, encompassing a variety of aspects and attributes of the outcome that are relevant to the individual.

Further, a risk issue is not necessarily perceived, as it is often presented, in isolation. Individuals perceive risks in relation to their wider beliefs about the risk issues and the more general implications that these have in their lives. For instance, beliefs about a risk issue may include both immediate and far-reaching consequences of accepting or not accepting the risk. And often the broader considerations will include probable benefits, which can be set against the risk in a risk–benefit trade-off. For example, suppose that a man is told by experts that the risk of death that he incurs by flying to his holiday resort is considerably less than that of driving by car on congested, foreign motorways. To understand his perception of the two risks, and thus the reason for his eventual choice of holiday travel, it is necessary to appreciate the systematic biases and heuristics that he may use to make his judgement. But before this can be done one has to grasp the *substance* of what he takes into account when thinking about the choice, i.e. how he defines the risk options and how they fit into his wider belief systems. The traveller may perceive the risks of these two modes in terms that are more differentiated than mere probability of death. He may take into account factors such as type of death and whether or not he has any control over the outcome. He may, for instance, perceive the outcomes of air travel as being in the hands of the pilot, whereas the outcomes of car travel are his own responsibility. Beliefs such as these may well affect how he feels about the two risks. Alternatively, he may accept the probabilities but choose the more risky route because he feels strongly about, or pays more attention to, the convenience of car travel. Thus his perception of the two risks is coloured by an implicit risk–benefit trade-off.

The research described here is based on this differentiated view of the outcomes of risk issues, in this case the generation of energy by various means. The complexity of risk perception is illustrated by examing what the public believes about different energy systems, the underlying dimensions that guide their thinking about the risk issues, and the feelings that they associate with different kinds of outcomes.

(*b*) *A convergence with attitude theory*

The approach that I have just described treats the perception of risk in the same way as that of any other issue: the construction by an individual of an internal model of the world primarily to plan and survive in that world. This internal model will have an information base derived from direct experience and information supplied by others and the media. It will also be a function of inference and processes of selective attention. However, perception and eventual action is not only guided by information but by feelings. By describing risk perception in terms of beliefs (the information base), feelings and attentional phenomena, all of which are seen as underlying action, I have pointed to a substantial convergence with attitude theory. This area of psychology can provide a conceptual framework and a methodology for studying the ways that individuals view the world, including its risks.

The particular expectancy-value model that underlies the research described in this paper is directly relevant to the conventional treatment of risk in that it has a probabilistic approach to the informational basis of attitude. *Attitude* is defined as an overall feeling of favourability or unfavourability toward an object or event. A *belief* is defined as an individual's endorsement of an association between an attitude object and an attribute (or consequence), as for example in 'the use of coal leads to air pollution'. The *strength of a belief* is the probability that the individual assigns to the association, i.e. his subjective probability judgement that 'the use of coal' will indeed lead to 'air pollution'. Each belief implies, in direct proportion to the strength with which it is held, some attribute (consequence) which evokes in the individual a feeling of favourability or unfavourability. For clarity this subsidiary attitude is referred to as an *attribute evaluation.*

At any given time, an individual's attitude toward an object can be treated as a summary of his attribute (consequence) evaluations for those of his beliefs that are within his span of attention. These are called his salient beliefs. Beliefs and attribute evaluations defined in this manner have been shown to relate to overall attitude in a relatively simple way (Fishbein 1963; Fishbein & Ajzen 1975):

$$A_j \approx \sum_{i=1}^{n} b_{ij} e_i, \tag{1}$$

where A_j is an individual's attitude toward object j; b_{ij} is the strength of his belief which links the attitude object j to attribute i; e_i is his evaluation of attribute i; $i = 1, 2, \ldots, n$ is the number of his salient beliefs, i.e. those currently within his span of attention.

This model of the relation between beliefs and attitude has been tested in a variety of different practical contexts (see Ajzen & Fishbein 1980); and it provides a method for describing and comparing belief systems and their associated affect. In this paper the emphasis will be on the belief aspect of risk perception; however, for nuclear energy, overall attitude and the values associated with different dimensions of risk will be discussed. The term *value* is used here specifically to refer to the average evaluation of the attributes (consequences) associated by an individual with a *dimension* of risk.

(c) *Comparative risk perception: some practical issues*

I have assumed that discussion and empirical study of risk perception reflects a concern with the social acceptability of risk standards. If this assumption is warranted, policy makers and agencies for risk regulation require knowledge of the relevant public's risk perception. The view presented here is that this implies more than a simple head-count of those in favour of or against acceptance of a particular risk; it requires an attempt to understand the beliefs and values that underlie the public's attitudes and their perception of risk.

If we accept that risk perception is an individual's response in a complex world then, strictly, an understanding of how risk is perceived requires a study of each individual as a single case. But on practical as well as epistemological grounds some form of aggregation is necessary. There are two sides to this question of aggregation. First, there are the practical difficulties of obtaining subjective reports on beliefs and feelings and converting these into equivalent categories that approximate to objective data, thus permitting comparisons between groups of individuals. To some extent this can be overcome by the method adopted here, i.e. by using verbal reports from depth interviews as a basis for choosing the most commonly occurring beliefs on which to establish a comparison.

Secondly, there are the theoretical difficulties that surround a choice of aggregation criteria. It is possible, by using the rich data that this method provides, to begin at the level of individuals and cluster them in terms of their risk perception on the basis of their beliefs, belief strengths or attribute evaluations. Alternatively, and this is the course adopted here, socially relevant groups can be defined *a priori*, and their perceptions compared across one or more risk issues. Given the controversy surrounding the nuclear energy programme in Austria at the time of the survey, we chose to examine not only the risk perception of the public sample as a whole, but also the differential perceptions of those who were most in favour of and most against the use of nuclear energy.

However, even given the will to understand risk perception in the sense used here, it is unlikely that resources would be available to mount an empirical study each time a risk issue is relevant to a policy decision. But it is possible that on controversial issues that have been debated in public (as was nuclear energy in Austria at the time of this survey), policy makers might possess sufficient information to understand the reasoning of public groups. Therefore, at the end of the survey we tested this hypothesis by undertaking a study of a group of policy makers. They were asked to infer the belief strengths and attribute evaluations that might underlie the risk perception of those in favour of and against the use of nuclear energy. The policy makers' accuracy was judged by comparing the public's actual scores with those that the policy makers attributed to them.

2. The survey of beliefs about energy system risks

(*a*) *Background to the survey*

The survey was directed toward a specific policy issue, the choice of energy options, and in particular the role of nuclear energy in the Austrian economy. During the research, the Austrian nuclear programme became an issue of considerable importance. As the first nuclear plant approached completion, the government organized a public information campaign, which began in late 1976. The progress of this compaign has been described in detail by Hirsch (1977). During the campaign, controversy built up to such an extent that a public referendum was held in November 1978. The referendum measure was narrowly defeated and the Zwentendorf nuclear plant, near Vienna, has not been brought into service. The survey about energy system risks was carried out at the end of the public information campaign; the study of the accuracy of policy makers' attributions to the public subgroups in favour of and against nuclear energy took place at the beginning of 1978, the year of the referendum.

(*b*) *The sample*

Representative sampling of the Austrian public was not possible, owing to financial constraints. Instead, a stratified sample was selected, which controlled for location (Vienna, provincial capital, rural), sex, age and education. The final

TABLE 1. BELIEFS ABOUT ENERGY SYSTEMS AND UNDERLYING FACTOR STRUCTURES

belief	factor structure†	
good economic value	F1	FII
increased standard of living	F1	FII
increased employment	F1	—
the industrial way of life	F1	—
increasing Austrian economic development	F1	FII
air pollution	F2	FIV
water pollution	F2	FIV
production of noxious waste	F2	FIII
making Austria dependent on other countries	F2	FIV
exhausting our natural resources	F2	—
changes in man's genetic make-up	F3	—
increasing rate of mortality	F3	—
(not) a technology that I can understand	F3	—
formation of extremist groups	F3	—
a police state	F3	—
new forms of industrial development	F4	FII
new methods in medical treatment	F4	—
dependency on small groups of experts	F4	FIII
technical spin-offs	F4	—
(not) exhausting natural resources	F4	FIV (R)
accidents that affect large numbers of people	F5	FI
exposure to risk that I cannot control	F5	FI
rigorous physical security measures	F5	FIII
hazards caused by human failure	F5	—
hazards caused by material failure	F5	—
exposure to risk without my consent	—	FI
a threat to mankind	—	FI
risky	—	FI
delayed effect on health	—	—
increases my nation's prestige	—	FII
reduces the need to conserve energy	—	—
satisfies the energy need in the decades ahead	—	—
long-term modification of the climate	—	—
decreases dependence on fossil fuels	—	—
increases the extent of consumer orientation	—	—
diffusion of knowledge about construction of weapons	—	FIII
transporting dangerous substances	—	FIII
destructive misuse of technology by terrorists	—	—
gives political power to big industrial enterprises	—	—
increases occupational accidents	—	FIV
physical risk	—	—

† Factors for all energy systems: F1, economic benefit; F2, environmental risk; F3, indirect (future-oriented) risk; F4, technological development; F5, psychological and physical risk. Factors for nuclear energy: FI, psychological risk; FII, economic and technical benefits; FIII, socio-political risk; FIV, environmental and physical risk. R, reversed item.

usable sample was 224 cases, although the number varies slightly across analyses where parts of the questionnaire data were incomplete. The breakdown of the sample across sampling categories showed a slight bias in favour of Vienna, males, youth, and high levels of education.

(c) *The questionnaire*

The beliefs about energy systems used in the questionnaire were chosen on the basis of open-ended interviews with members of the Austrian public, from a pilot survey of beliefs and attitudes of energy experts toward nuclear energy (Otway & Fishbein 1976) and a literature survey. The complete set of 39 belief statements used in the questionnaire is shown in table 1. The respondents were asked to give demographic information, and to scale their attitudes toward all five energy systems and toward the 39 attributes (consequences) associated with the set of belief statements. They were then asked to rate the 'truth' of each statement related in turn to each of the energy systems. The scaling methods used were as follows.

(i) *Overall attitude* was measured by using the semantic differential technique (Osgood *et al.* 1957). The attitude object was rated on a series of seven-point scales (+3 to −3), where the endpoints were labelled with adjective pairs such as 'good/bad'. In keeping with Osgood's procedure, a factor analysis of the responses was used to identify those adjective pairs that loaded on an evaluative dimension; five pairs of adjectives were validated in this way and the measure of overall attitude was a sum of the ratings on these five scales, giving a range of +15 to −15.

(ii) *Attribute evaluations*, i.e. attitudes toward the attributes (consequences) associated with each belief, were measured in a manner similar to overall attitude, but with a single evaluative scale labelled good/bad.

(iii) *Belief strengths* were measured by using seven point scales (+3 to −3) where the endpoints were labelled likely/unlikely. It should be noted that although belief strength has been construed as a subjective probability, in keeping with most of the research based on Fishbein's model it has been scaled in a way that avoids certain requirements of probability measures. The beliefs are not treated as a partitioned event space where the probabilities would sum to unity, and the use of a bipolar scale made it possible to encompass levels of probability that the energy system *is* or *is not* associated with the attribute (consequence) in question.

All questionnaires were administered in a personal interview, which permitted full instruction on the use of the scales.

3. Results

(a) *Attitudes toward five energy systems*

Although the primary concern of this report is a comparison of the beliefs that underlie risk perception, it is worth placing this in context by first considering attitudes toward the various means of energy generation.

The attitude scores (measured by the semantic differential technique) for the total sample yielded the three distinct types of frequency distribution shown (smoothed) in figure 1. The distributions were virtually the same for the two fossil fuels, as were those for hydro and solar energy. The distribution for nuclear energy was quite different. For fossil fuels there were very few negative and few highly positive attitudes; most respondents were moderately favourable. For hydro and

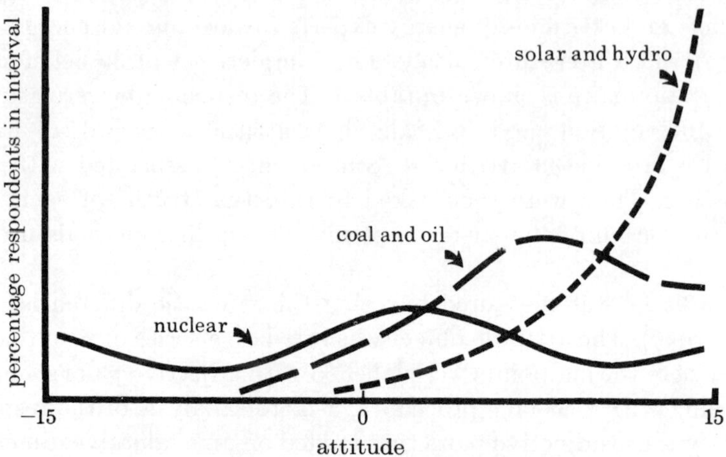

FIGURE 1. Smoothed frequency distribution of attitudes towards energy systems. Sample of Austrian public, $n = 211$.

solar energy there were virtually no negative attitudes. The most frequent response was highly favourable. Attitudes toward nuclear energy were most frequently near the middle of the scale, but there were also clusters of highly positive and highly negative attitudes. Only for nuclear energy were the attitudes sufficiently polarized to warrant differential analysis of underlying beliefs for those most in favour of and those most against the use of nuclear energy. Two subgroups of the sample were formed by selecting the 50 respondents with the most positive attitudes toward the use of nuclear energy (pro group) and the 50 most against its use (con group).

(b) *Risks and benefits: the underlying structure of beliefs about energy systems*

To identify the commonalities in perceptions of the energy systems, the 39 beliefs used in the questionnaire were reduced to a smaller number of underlying dimensions by factor analysis. Factor analysis is a generic term for several linear, parametric statistical methods that identify the minimum number of dimensions required to account for the variance in a larger set of intercorrelated variables (here the ratings of belief strength). Our first concern was how many generalized dimensions of belief would represent the 39 attributes (consequences) when each was related to all five energy systems in turn. Tucker's extension of the factor analytic procedure to three-dimensional matrices was used to simplify the data cube n by m by q, where n subjects had responded to m belief statements about q energy systems (Tucker 1966). The method used was based on a three-way decomposition of the raw cross-products matrix, followed by DAPPFR rotation. This produces oblique (correlated) factors, although in this case the intercorrelations between the factors were low. Full details of the findings for all three modes, subjects, beliefs and energy systems can be found elsewhere (Thomas *et al.* 1980*a*). Only the belief mode is discussed here.

Five belief dimensions were found to underlie the public's perception of all the energy systems taken together. Three of these were dimensions of risk and two were concerned with probable benefits. The risk dimensions were environmental risk, indirect risk (future-oriented and political), and psychological and physical risk. On the benefits side the two dimensions were general economic benefits, and future-oriented benefits of technological development. The five belief items most closely associated with each factor can be seen in table 1.

A second factor analysis was performed, by using the belief strength ratings for nuclear energy alone. The method in this case was that of principal components analysis, followed by Varimax rotation. This technique (unlike the three-mode method described above) produces underlying dimensions that are independent, i.e. orthogonal factors. For nuclear energy, a set of belief dimensions emerged that were similar to those for all five energy systems but different in some important details. The indirect risk factor was redefined as a more obviously socio-political risk dimension. There was a psychological risk factor, while the beliefs about physical hazards now clustered with environmental risk on a dimension which was called environmental and physical risk. For nuclear energy there was a single benefits factor that combined economic and technical benefits. The five beliefs most closely associated with each of the four dimensions are indicated in table 1.

(c) *Comparative perception of risks and probable benefits of energy systems*

Perceptions of risks and probable benefits were compared across all five energy systems for the pro and con nuclear energy subgroups of the public sample. For every individual in the pro and con subgroups a factor summary score was calculated by summing the belief strength scores of the five most characteristic beliefs associated with each belief dimension of the five-factor solution. Mean values of these factor summary scores are shown for the pro and con subgroups as bar-diagrams in figure 2. An inspection of figure 2 shows the following points.

(i) The energy systems appear to be characterized by profiles of varying belief strength about risks and probable benefits.

(ii) Apart from nuclear energy, across all the energy systems there is an overall dissociation with risks. This is particularly strong for hydro and solar energy. The only exceptions are the beliefs that fossil fuels, especially oil, are associated with environmental risk.

(iii) There is an overall association of the energy systems with probable benefits. The most notable exception to this is a belief that coal will not lead to technological development. The con nuclear group also believed that hydro energy and oil will not lead to technological development and that coal will not lead to economic benefits.

(iv) Across all the energy systems, the pro nuclear group believed more strongly in the economic and technological benefits of energy generation than did the con group.

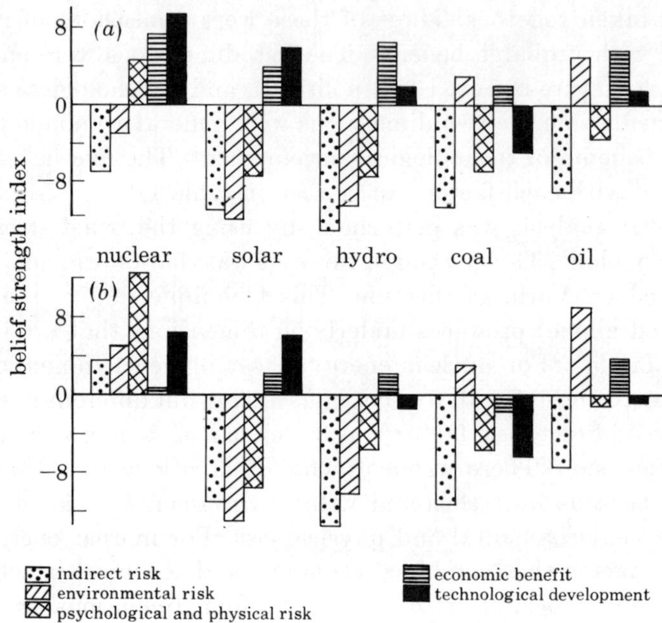

FIGURE 2. Beliefs about five energy systems held by those pro (*a*) and con (*b*) the use of nuclear energy. Subgroups of Austrian public sample: pro nuclear group, $n = 48$; con nuclear group, $n = 47$.

(v) The perception of nuclear energy differs from the other energy systems in that it is the only one associated with indirect risk (con nuclear group), and psychological and physical risk (both con and pro groups).

(vi) When the perceptions of those pro and con nuclear energy are compared across the energy systems, apart from their perceptions of nuclear energy they are strikingly similar.

(vii) The pro and con nuclear energy groups perceive the risks and probable benefits of nuclear energy very differently. (These differences are statistically significant for all five dimensions.)

(viii) Perception of the risks of nuclear energy: the con nuclear group believed that the use of nuclear energy would lead to all three dimensions of risk, the strongest belief being in psychological and physical risk. The pro group associated nuclear energy with only one risk dimension, that of psychological and physical risk, but this belief was less strong than for the con group.

(ix) Perception of the probable benefits of nuclear energy: the pro nuclear group believed strongly in the economic and technological benefits of nuclear energy. The con group also believed that nuclear energy would lead to technological development, but they had only very weak belief in its economic benefits.

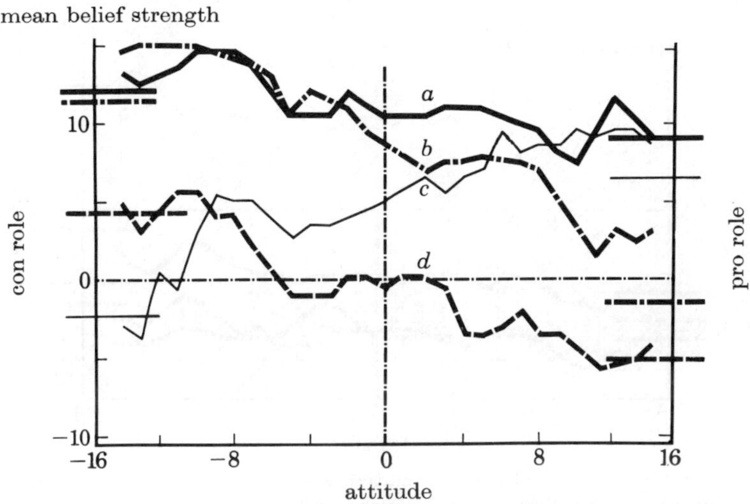

FIGURE 3. Relation between strength of belief on each dimension and attitude towards nuclear energy. Sample of Austrian public, $n = 224$: pro role policy makers, $n = 17$; con role policy makers, $n = 18$. (*a*) Socio-political risk; (*b*) psychological risk; (*c*) economic benefit; (*d*) environmental and physical risk. Horizontal bars are the mean values attributed to the public by policy makers in pro/con roles.

(*d*) *Nuclear energy: risk perception and attitude*

Since the pro and con groups were defined in terms of their overall attitudes toward nuclear energy, it is worth stressing that perceptions of the risks and probable benefits contributed significantly to their attitudes. A Pearson correlation coefficient of 0.66 was found for the relation between a direct measure of attitude by using the semantic differential technique and a belief-based measure of attitude calculated according to equation (1) and by using factor summary scores derived from the four factor solution. Details of this validation can be found in Thomas *et al.* (1980*b*, *c*). The relation between beliefs in the risk and benefit dimensions and attitude to nuclear energy are shown in figure 3 for the entire Austrian sample. In this figure, factor summary scores for belief strength are plotted as a series of lagged means based on the scores of those individuals whose attitudes fall at each point on the attitude scale. As might be expected, when attitude increases (from negative through neutral to positive), the perceived association with the three risk dimensions decreases, whereas belief in the benefits of nuclear energy increases. It is worth noting that the belief in socio-political risk changes rather less with attitude than do the other risk dimensions.

So far this paper has focused on comparisons of the belief aspects of risk perception. But for nuclear energy there is strong evidence that these beliefs contribute to the respondents' attitudes, and it is therefore worth looking also at their feelings about the various probable outcomes. Factor summary scores for attribute evaluations were calculated in the same way as for belief strength, by using

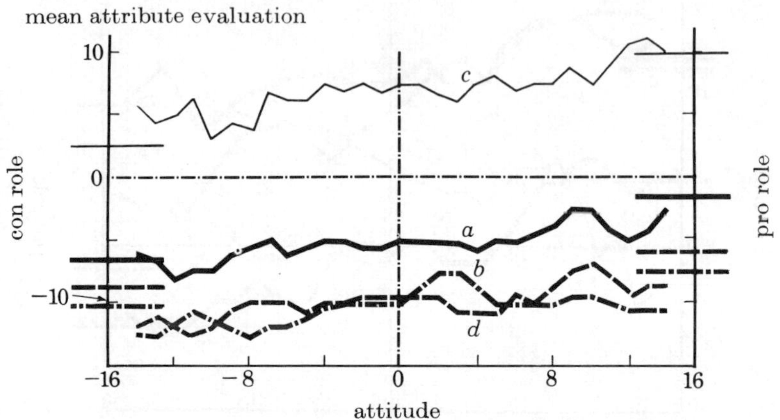

FIGURE 4. Relation between feelings associated with each belief dimension and attitude towards nuclear energy. Sample of Austrian public, $n = 224$: pro role policy makers, $n = 17$; con role policy makers, $n = 18$. Symbols as in figure 3.

the five most characteristic beliefs on each dimension. In the case of attribute evaluations these factor summaries have been defined as the *values* associated with each belief dimension. Changes in the four values with attitude toward nuclear energy are shown in figure 4. It can be seen from figure 4 that as the respondents' attitudes become more positive the positive value of economic benefit also increases, and the negative value associated with the three risk dimensions becomes less strong. It should be noted that although these changes do not appear large, the differences between the pro and con groups were statistically significant for all four dimensions; both the respondents' values and their belief strengths contributed to the observed differences in attitude.

(e) *The accuracy of policy makers' attribution of beliefs and values to the pro and con nuclear energy subgroups*

The results so far presented give some indication of the complexity of risk perception; the next step was to find out whether our group of policy makers could, without the advantage of the kind of data described here, reproduce the complexity of the public's risk perception. In particular we were concerned to find out whether, in the period between the public information campaign and the referendum, the policy makers could infer the differential risk perceptions of those pro and con nuclear energy.

We were fortunate to have the full cooperation of the Energy Section of the Austrian Federal Ministry of Trade and Commerce. The sample of policy makers were 40 senior civil servants who advised the government on energy matters. Each participant was assigned (randomly) to one of two role-play conditions and asked to complete the same questionnaire as the public but in the role of a typical Austrian citizen who was in favour of or against the use of nuclear energy. These instructions were intended to discourage the role-playing of militant extremists. The data were analysed in the same way as for the public sample and factor summaries for belief strength scores and attribute evaluations were calculated for the same four belief dimensions.

Since the instructions to the policy makers only loosely defined pro and con subgroups of the public, rather than making a direct comparison between the policy makers' in-role responses and the public pro and con subgroups as previously defined, the in-role data, averaged within each role-play condition, have been compared with the beliefs and values of the entire public sample. The mean factor summary scores for the policy makers' in-role responses have been added, in the form of horizontal bars, to figures 3 and 4, representing their attributions of belief strength and value respectively. This method of presentation gives an immediate picture of the overall accuracy of the policy makers' attributions to the pro and con subgroups.

Despite the general accuracy, however, there were discrepancies and these form a consistent pattern. Regarding belief strengths the policy makers were inaccurate on the psychological risk and economic and technical benefits dimensions. For both the pro and con subgroups the policy makers underestimated their belief that nuclear energy is associated with psychological risk. The misperception of the pro subgroup's belief on this dimension was the largest and most striking error of attribution in the role-play experiment. On the economic and technical benefits dimension, the misperception was again in the same direction for pro and con subgroups, this time an underestimate of the strength of belief that nuclear energy leads to these benefits.

The policy makers' attribution of values was, once more, inaccurate on the same dimensions and in the same direction for the pro and con subgroups. They underestimated the negative value that both groups associated with psychological risk and with environmental and physical risk. They also underestimated, but to a lesser degree, the pro subgroup's negative feelings about socio-political risk.

4. Discussion

This paper has presented a view of risk perception that treats it essentially as an instance of perception in general. Thus risk perception can be seen as a function of the way an individual views the world, his beliefs, feelings and plans for survival in that world. We have shown that by adopting an expectancy-value model of attitudes and beliefs and the techniques appropriate to that model, it is possible to measure belief strengths and feelings in relation to risk issues, and to use these measurements as a means to understanding how the public perceives risk.

The most important general finding of the study is that the public does indeed appear to conceive risk issues in differentiated terms, taking into account several substantive dimensions, which include probable benefits as well as unpleasant outcomes. Two similar sets of dimensions were identified as underlying the perception of five energy systems taken together (coal, oil, hydro, solar and nuclear energy) and nuclear energy considered alone. For all five energy systems there were three dimensions of risk and two of probable benefits. For nuclear energy the two benefit dimensions coalesced, and the three risk dimensions differed somewhat in detail.

What the public perceives as relevant dimensions of risk and benefit will vary with the risk issue in question and with the extent of public debate and public knowledge. But what does seem likely is that the public will structure their thinking on more dimensions than usually receive attention in the traditional equations of experts. For example, the Austrian public information campaign initially focused on medical concerns. But pressure groups, and to a lesser extent public opinion, forced a change in the content of the debates, introducing other aspects of risk such as the socio-political implications of nuclear energy. The nature of the risk and benefit dimensions found here (indirect risk, psychological risk, socio-political risk, physical risk, environmental risk, general economic benefit and technological development) could well be relevant to technological risks in general. But, at present, this is an empirical question.

On the other hand, the possibility that the perception of risk issues always depends on dimensions of risk *and* probable benefits, an implicit trade-off procedure, has a theoretical basis in attitude theory. An individual's risk perception can be considered as instrumental in making some pro/con judgement about acceptability i.e. forming an attitude to the risk issue. And the assimilation of favourable and unfavourable, or pro and con attributes (consequences) seems to be fundamental in forming attitudes and making choices.

This paper has shown not only that the public's risk perception is multidimensional, but that the public, and different subgroups within the public, can distinguish between risk issues (in this case energy generation systems) in terms of belief strengths and feelings about each risk or benefit dimension. But if the differentiated view of risk perception that has been put forward here is to be used in the service of policy making, then policy makers and risk regulation agencies will need a fairly sophisticated understanding of the public's points of view.

We have shown that policy makers, at least on issues that have aroused public concern and debate, are able to infer the belief systems and values of prominent public groupings with quite a large degree of accuracy. But one of the more important findings of this research is that, despite the public controversy surrounding nuclear energy in Austria, the policy makers were strikingly wrong on certain dimensions, particularly on the public's concern with the psychological risks of nuclear energy. Their mistaken attribution was due both to an underestimation of the negative feelings associated with the outcomes of the psychological risk dimension and to an underestimation of the high probability assigned to the relation between use of nuclear energy and these unpleasant psychological outcomes. Both of these inaccuracies were especially large for the pro nuclear subgroup; and if such errors in attribution were widespread in Austrian government circles, they could well have contributed to the Government's overoptimistic stand on the Zwentendorf referendum.

Finally, it should be stressed that a subjective view of risk will inevitably imply an attempt to understand risk perception at the level of different groups in society. And no matter how much research is done, or how informed and accurate in their attributions the policy makers and those charged with setting risk standards may be, the problem of how to reconcile conflicting risk perceptions remains.

The research on which this paper is based was carried out as part of the Joint Risk Project of the International Institute for Applied Systems Analysis and the International Atomic Energy Agency, Vienna, Austria. The data were collected and analysed by a team of researchers, and the author would like to acknowledge the assistance and advice of Martin Fishbein, Ron Hinkle, Dagmar Maurer, Harry Otway, David Simpson and Elizabeth Swaton. The research team would also like to thank the members of the Energy Section of the Austrian Federal Ministry of Trade and Commerce who took part in the role-play experiment.

References (Thomas)

Ajzen, I. & Fishbein, M. 1980 *Understanding attitudes and predicting social behaviour.* Englewood Cliffs, N.J.: Prentice-Hall.

Fishbein, M. 1963 *Hum. Relat.* **16**, 233–240.

Fishbein, M. & Ajzen, I. 1975 *Belief, attitude, intention and behavior: an introduction to theory and research.* Reading, Mass.: Addison-Wesley.

Hirsch, H. 1977 Presented at the International Conference on Nuclear Power and its Fuel Cycle, Salzburg, Austria.

Kahneman, D. & Tversky, A. 1973 *Psychol. Rev.* **80**, 237–251.

Osgood, C. E., Suci, G. J. & Tannenbaum, P. H. 1957 *The measurement of meaning.* Urbana, Illinois: University of Illinois Press.

Otway, H. J. & Fishbein, M. 1976 *The determinants of attitude formation: an application to nuclear power*, RM-76-80. Laxenburg, Austria: International Institute for Applied Systems Analysis.

Thomas, K., Maurer, D., Fishbein, M., Otway, H. J., Hinkle, R. & Simpson, D. 1980*a* *A comparative study of public beliefs about five energy systems*, RR-80-15. Laxenburg, Austria: International Institute for Applied Systems Analysis.

Thomas, K., Swaton, E., Fishbein, M. & Otway, H. J. 1980*b* *Nuclear energy: the accuracy of policy makers' perceptions of public beliefs*, RR-80-18. Laxenburg, Austria: International Institute for Applied Systems Analysis.

Thomas, K., Swaton, E., Fishbein, M. & Otway, H. J. 1980*c* *Behav. Sci.* **25**, 332–344. (Reprint of Thomas *et al.* (1980*b*).)

Tucker, R. L. 1966 *Psychometrika* **31**, 279–311.
Tversky, A. 1974 Presented at a colloquium on theories of decision and their semantics, Salzburg, Austria.
Tversky, A. & Kahneman, D. 1974 *Science, N.Y.* **185**, 1124–1131.

Discussion

H. J. Dunster (*Health and Safety Executive, London, U.K.*). I was impressed by the accuracy of the Austrian administration. Even in the worst area, I thought they did remarkably well. I hope that the corresponding people in this country would do as well.

Kerry Thomas. While I agree that the Austrian policy makers were largely accurate in their attribution to the public, and we were ourselves a little surprised by the findings, nevertheless there were inaccuracies that were not only statistically significant but which showed clear misunderstanding of the pro nuclear public. I am referring to the policy makers' attribution to the pro public of a disbelief in the psychological risks of nuclear energy. In fact the public as a whole believed that such risks are associated with nuclear energy.

C. Green (*School of Architecture, Dundee, U.K.*). An undefined area in the Fishbein attitude model is the question of salient beliefs: what they are, why they are and where they come from. His reference to Miller's Magic Number 7 ± 2 seems to me essentially an easy way out. Does Dr Thomas see salient beliefs as referring to values?

Also, while Dr Thomas has been operating with the A-object attitude model rather than a multi-attribute utility model, does she think that examining choice between alternative energy strategies may not be more useful in the long run than exploring attitudes towards energy systems?

Kerry Thomas. There is no doubt that the concept of salient beliefs is problematic, not least since the level of their generality determines the information that they encompass. I would not want to identify salient beliefs with values, but clearly a person's values will influence which areas of information on a given topic are normally taken into account. And the more general the beliefs, the more likely it is that one might be able to infer the person's values.

On the second point, I agree that beliefs about choices between energy alternatives would provide more information than beliefs about energy systems taken one by one. However, the methods associated with choice models are not easily adapted to survey research. I would hope that a decision analysis approach might be used as a next step in this research area.

Proc. R. Soc. Lond. A **376**, 51–64 (1981)
Printed in Great Britain

QUANTIFICATION OF BIOLOGICAL RISK

Assessment and evaluation of risks to health from chemicals

By A. E. M. McLean

Laboratory of Toxicology and Pharmacokinetics, Department of Clinical Pharmacology, University College Hospital Medical School, University Street, London WC1E 6JJ, U.K.

Asbestos, thalidomide, and smog in the environment have all given spectacular evidence of the power of man-made chemical substances to harm people. Phenobarbitone, paracetamol, DDT and penicillin are chemicals that have given large benefits for small risk. Epidemiological evidence allows us to consider dose, response and cost for some of these.

For new chemicals we try to assess risk before human exposure starts, using model test systems. The tests are based on studies of substances known to have caused harm to man. The methods of risk and benefit assessment are not perfect, so that post-exposure surveillance is necessary for new and old drugs, pesticides and industrial and environmental chemicals.

Risk assessment then takes the form of a continuous review of chemical use, as the relative risks, costs, benefits and alternatives change with developing technology. Those at the exposure end may have interests that conflict with the interests of the rest of society, so that any evaluation process for the risks should include a process for winning the agreement of the people at risk.

The case of 2,4,5-T shows how the scientifically difficult and socially fragile process of risk assessment and evaluation can easily be disrupted. We need more systematic processes of quantification and more robust processes for evaluation of risks from chemicals.

Introduction

There have always been poisons, and the acute effects of plants and toxic fungi like the Amanitas have been with man since he was a hunter–gatherer. For those who lived long enough there was liver cancer from aflatoxins in mould-contaminated food, and breast cancer associated with a high fat intake. But as men started to mine the ground and to synthesize new chemicals, the variety of highly toxic materials and the number of men exposed to them increased (Hunter 1975). There are now perhaps a thousand new chemical substances marketed every year and ten thousand sold at more than 500 t annually. A quarter of a million natural substances of low molecular mass are known. What should and what can we do about all these (Zbinden 1980; Korte 1977)?

Health in this country is better than in past times. The expectation of life in this chemical era is longer than ever before. Infant mortality is far lower than it was 50 years ago (D.H.S.S. 1980). However, we see from table 1 that the ratio of infant mortality between poorer and more prosperous groups in the society is the same now as it was in the 1930s. One way of saying this is that the poor have shared equally in improvement of health with the prosperous; 61 % in both social class I and social class V. The absolute figures suggest that social and material deprivation increase infant mortality. Poverty is also bad for the health of adults, particularly in the diseases caused by cigarette smoking. The diseases associated with accidents at work and accidents in the home are also powerfully concentrated in poorer groups. As a result, teachers have a much longer life expectancy than warehousemen (McLean 1979). Cigarette smoking (table 2) is heavily concentrated in the poorer groups and one can discuss why this is and how it comes about, but the fact and its effects are incontrovertible.

Table 1. Trends in infant mortality, by occupational class

(from D.H.S.S. (1980).)

	deaths per 1000 live births		percentage decrease
	1930–2	1970–2	
class I	32	12	62
class V	80	31	61
ratio, V/I	2.50	2.59	

Table 2. Cigarette smoking and occupation

(Percentages of men that are current smokers. From O.P.C.S. (1980, table 8.13).)

Professional	25
Employers and Managers	37
Intermediate non-manual	38
Skilled manual	49
Semi-skilled manual and personal service	53
Unskilled manual	60

For the cancers, age-specific incidence is fairly static except for the large increase in lung cancer over the last 40 years. For stomach cancer and for some others, incidence is going down. We see a population living in a chemical era healthier than ever before. There is no evidence of mass harm to the population from new chemicals (Doll 1979; D.H.S.S. 1980).

Areas of concern: industrial chemicals

One has only to look at any issue of the *British Journal of Industrial Medicine* or any of the other journals of industrial medicine to find reports of ill-health from exposure to solvents, to carbon disulphide (with increase in coronary heart deaths), dust (leading to diseases of the lungs), and many other problems of health in industry. Some are associated with particular chemicals, others associated with dusty, cold or wet working conditions (see Hunter (1975, p. 1144)).

Chemical injuries are not the most numerous of the industrial accidents. However, they are an important and sometimes fatal matter: bladder cancer was a major hazard for workers in the dye and rubber industry until β-napthylamine was banned (Williams 1958). It is notable that where measures of exposure to carcinogenic materials are available to us, we find that as dose goes up so the proportion of men getting cancer goes up, and as exposure goes down, so the proportion falls, and we can predict that, if the levels of exposure go down, so the proportion of men developing tumours will go down to very low levels. We predict that if we can control the levels, tumour incidence will be pushed down to 1 in 10^6 1 in 10^7, and so on even after a lifetime of exposure. It is of course extremely difficult to measure such small risks.

Past asbestos exposure is causing perhaps 200 deaths a year in the U.K. from mesothelioma alone, and perhaps another 1000 a year from lung cancer from combined cigarette and asbestos exposure. These figures are dreadful even in the context of a background of around 290000 deaths per year, of which 59000 are cancers, and 23000 lung cancers among men in England and Wales (O.P.C.S. 1978; Registrar General 1967).

However, the estimates that up to 40% of all cancers are due to industrial chemicals are totally baseless. They are derived by multiplying the risks for men heavily exposed to carcinogenic chemicals by the total number of persons ever exposed to the much lower levels found in present conditions (Peto 1980; Heavey *et al.* 1980).

Medicines

The side effects of drugs are a common hazard. Perhaps 1 in 10 of our older patients in University College Hospital has a recognizable side effect, which can vary from constipation to a rash or renal failure, all of them serious matters in old people (Connell 1979).

Household chemicals and pesticides

Household chemicals such as cleaning fluids and the pesticides cleared for home and garden use have proved immensely safe. Even the attempts of suicide with such materials usually come to nothing so long as strong acids or alkalis are not used. There are occasional reports of the use of agricultural pesticides causing illness, but these are usually relatively minor side effects of misuse or mismanagement. It is many years since anyone died from accidental exposure to pesticides, though suicide with agricultural pesticides such as paraquat is well known.

Regulation of chemical use

What shall we do about all these old and new chemicals? We fear the introduction of a new substance as toxic as the cigarette. The products that we have to think about can be classified as drugs, pesticides, food additives and household chemicals, and particularly industrial chemicals. Because I have done experimental work and epidemiological work on toxicity of drugs and industrial solvents, I have been involved in the committees concerned with regulation of chemicals in food, in the regulation of pesticide use, and in mutagenesis (table 3).

Table 3. Main procedures for measurement of pesticide toxicity

1. single dose	acute effects, target tissue; $\mathrm{l.d.}_{50}$, several species
2. 90-day feeding	observation, chemistry of tissues and histology
3. neurotoxicity, teratogenicity	chickens, two species, often multi-generation reproduction
4. mutagenicity	bacteria and mammalian
5. carcinogenicity	2 years rat or mouse, with long-term observation and histology
6. skin, eye and inhalation	irritancy and damage by contact
7. metabolism	in animals and plants; absorption, and routes of removal, and toxicity of metabolites
8. wildlife	birds, fish, insects, plants, water
9. human reactions	allergy or other effects in workers in manufacture or use

What do the advisory committees do (M.A.F.F. 1979)? Essentially, they can look at the evidence produced by the manufacturer. For a pesticide perhaps £2M or £3M worth of work is presented, more than a hundred man years of work and many thousands of animals sacrificed. We look first at the effect of a single dose and see what size of dose is required to cause harm and to cause death. We are looking not only for the comparative toxicity between different substances, for the way in which different species respond, but also we are particularly interested in knowing the target organ or tissue which the substance attacks. We consider 90-day studies of repeated feeding. We look at mutation; we are particularly interested in evidence of carcinogenic effects. We are interested in the sensitivity of the skin and eye because one of the common things that happens with chemicals is that they may be splashed on the skin or eyes; we want to know if they cause irreversible damage or allergic reactions. We are interested in inhalation, for substances that are going to be sprayed. The effects on reproduction are of particular interest and importance because the production of malformed babies is an irreversible personal and social disaster. We are concerned with the mechanism by which toxic effects take place and the metabolism of the substance.

Most of these toxic effects are crucially dependent on the dose. All substances are harmful if the dose is large enough, as Paracelsus said 400 years ago. We try to define the target organ, the dose at which there is harm, and the dose at which there is no longer a harmful effect. Very often that depends not on the dose administered, say the dose that lands on the skin, but the dose that lands finally in the target tissue. Pharmacokinetic studies have become increasingly important in extrapolation of animal data to man.

The dose that produces harmful effects varies enormously from species to species and from laboratory to laboratory, depending on the diet and the strain of animal. For the organophosphorus insecticide carbophenothion there is a very wide variation in toxicity. The l.d.$_{50}$ varies from 7 to 1250 mg/kg (table 4).

TABLE 4. CARBOPHENOTHION l.d.$_{50}$ (MILLIGRAMS PER KILOGRAM BODY MASS)

(Stanley & Bunyan (1979).)

Rat	7
Rat	91
Mouse	106
Rabbit	1250
Starling	5.6

IMPORTANCE OF MOLECULAR MECHANISMS

If we know the molecular mechanism involved in toxicity and if we understand the pharmacokinetics (i.e. absorption, distribution, metabolism and excretion), we have a much better chance of finding not just the variation between toxic effects in different species but their common factors. We then have a better chance of predicting what will happen to man. For instance if an enzyme is needed for a particular toxic effect in animals, and we know that man has the enzyme, then we can expect man to be susceptible. For instance, in the liver from an animal treated with carbon tetrachloride there is a massive destruction of liver cells. When one alters the enzyme cytochrome P_{450} in the liver one alters the lethal dose and the amount of liver injury because the enzyme P_{450} is the rate-limiting step required to oxidize carbon tetrachloride into the real toxin, a highly reactive and destructive metabolite molecule (Garner & McLean 1969; McLean & Day 1975; Chenery & McLean 1980).

We know that man has the enzyme in both the liver and the kidney, and as a result we could predict that man is susceptible even had we not known from clinical studies that he was. So we can measure toxicity and relate it to dose, and then find doses for which the response gets less and less till no effect is to be seen.

CARCINOGENIC RISKS

For cancer-producing chemicals we have a different kind of change in effect when we change the dose. One does not get a little bit of cancer; one either has it or one does not. At lower doses, smaller and smaller proportions of an exposed population develop the tumour after a lifetime of exposure, or a single exposure. And as a rule we can say that as with acute effects, the dose that matters is not what lands on the skin or in the air, but the dose that lands in the target tissue.

We have good evidence that this is true not only for animals but also for man. For cigarettes there is a good dose–response curve (see Acheson, this symposium). There is no convincing evidence of effective threshold dose effects, so one is left with the concept of doses low enough to produce a risk that is not intolerable.

Environmental carcinogens and the 'safe dose'

One used to think that it might be possible to eliminate carcinogenic compounds from the environment, but it has become clear that there are so many naturally occurring substances in food and made in the body that can initiate the process of carcinogenesis that it is not a question of eliminating them, it is a question of detecting those that are of importance and trying to reduce their level to a point that we find acceptable. For instance, nitrosamines are present in bacon, they are made in our stomachs from nitrite and amines naturally present in our foodstuffs, and they may be made by bacteria in the gut. When dipropylnitrosamine was found in low concentrations as a contaminant in a very widely used herbicide, Trifluralin, there was a problem of deciding whether this posed a real risk. The United States Environmental Protection Agency (E.P.A.) produced a calculation of risk. The dose and the population exposed were estimated. Trifluralin is used to clear the ground for planting soya beans, cotton, sugar and a very large number of other crops. About 340000 people are exposed, and the dose calculated was about 1 μg per person per year. From animal data the E.P.A. derived an approximate equation of the risk against the dose during a rat's lifetime, and applied this to the human situation. The equation has a component for the length of exposure and the dose is expressed in terms of what it would be if the carcinogen were in the food as it was in the animal experiments. Using this calculation and hoping that the corrections will allow for the species variation between rat and man, E.P.A. have come up with figure for risk (table 5).

For 5 mg dipropylnitrosamine/kg in trifluralin a life-time cancer risk of 0.04 cases in the 340000 people exposed was calculated. This must be compared with the many thousands of tumours to be expected from the ordinary background of risk in the population studied (E.P.A. 1979).

I think that we can say that it is a negligible risk. The exposure dose is equivalent to eating a pound of bacon a year and it seems unrealistic to treat it much more seriously.

There is controversy about the correct equation to use in predicting the risk for people exposed to low doses of carcinogens, based on animal data, or information from human epidemiology where high doses have made the risk detectable (Hoel 1980; Gehring *et al.* 1979).

Many equations will fit the limited and uncertain data points that we have.

However, it seems reasonable to prefer calculations that take into account biological factors such as the finite time required for one initiated cell to develop into a tumour, the need for most carcinogens to be metabolized to a reactive molecule, and the pharmacokinetics of the compound, which determine dose to target tissue.

Gehring (1979) has performed a series of such calculations for the exposure of men to vinylchloride monomer (table 6). Polyvinylchloride (PVC) is an enormously important substance and it has great economic effects. We have better and cheaper drain pipes and so drier and warmer houses because of PVC.

Table 5. Life-time tumour risk from the herbicide trifluralin (E.P.A. (1979).)

(*a*) *equation for tumour risk*

$$I = 1 - \exp(-bdt^m),$$

where I = tumour incidence as proportion of population at risk; t = time, fraction of life; d = daily dose in diet (mg/kg diet); m = constant (for time); b = factor (all species).

(*b*) *example: contamination by* 5 mg *dipropylnitrosamine*/kg *diet*

U.S. number exposed	340000
time	40–70 years
annual dose	1 µg
m	3
b	0.4
risk	0.04 cancers per lifetime of population
background cancers	68000
annual loss to farmers if banned	$300 M

Table 6. Vinyl chloride and angiosarcoma

(Gehring *et al.* (1979).)

9677 men at risk; 5 angiosarcomas of liver found
Predicted: (*a*) 10 (probit model), (*b*) 2 (linear model), (*c*) 50 (linear, no threshold), (*d*) 53 (exponential (N.C.I.))

In the past there was severe exposure to the vinylchloride monomer in the manufacture of PVC and we have data about workers who developed angiosarcoma of the liver as a result. Looking at the rat data, Gehring pointed out that the dose–response relation was not simply one of how much vinylchloride monomer (VCM) there was in the air; what was needed was to know how much got to the liver and was metabolized there to the reactive metabolite. Using the figures for VCM metabolism and drug metabolism in rat and man, he was able to make a reasonably good dose–response equation for liver cancer in the rats. Using these reasonable dose constants for the men exposed at work, Gehring was able to test the predictive force of a number of different equations. The Environmental Protection Agency calculation quoted above rather overestimates the number of cancers that are expected in comparison with those that were found. These calculations have great uncertainties in the constants that we use, in the equation we use and in the measurement of dose. However, I think that they give us answers that are probably correct within two orders of magnitude. Probably the trifluralin risk is even less than is calculated by E.P.A., while one can suspect that there are going to be more tumours with time in Gehring's study population exposed to VCM in past years. But it is such calculations that enable us to make a fairly rational assessment of risk to people from the universally present initiating carcinogens in our environment.

Overestimates of risk from a chemical will lead to diversion of resources to reduce exposure, with no corresponding benefit. At the same time, the diversion of resources can lead to real deprivation through reduced provision of materials such as housing. This would have a detrimental effect on health. The calculation of the effect of putting increased resources into environmental or worker safety on social and material wellbeing for different social groups is immensely complex: the biological scientist can only point to the biological consequences of the various possible causes of action (E.P.A. 1979). Underestimates of risk have more directly adverse effects, but can still be difficult to detect.

Distribution of risks

How can we justify a single person getting liver cancer when others have the benefits? The same problem exists for men working as deep-sea fishermen. All social activities have various risks attached to them. The decision of whether any risk is justified by the benefit is relatively easy for medicines, where risk and benefit go to the same person, but it is of course extremely difficult for industrial chemicals. Fortunately, in the vast majority of instances there is no expectation of risk at all. For instance, for the antioxidants in food, all the animal evidence suggests that the human dose is one that produces no risk and may have a slight anti-cancer effect. For products like pesticides, we usually have an existing substance with which to compare the new product. As a result we are able to say 'this substance is acceptable because the risk it offers is less than the risk of the existing substance'. On this basis, for instance, the new synthetic pyrethroids gain, in comparison with organochlorine and organophosphorus compounds. With very widespread use of pyrethroids the dose will go up and one will begin to have doubts about the adequacy of animal data on their own.

It was on this kind of procedure that the Pesticides Safety Precautions Scheme (P.S.P.S.) rejected the soil fumigant dibromochloropropane (DBCP) because the Committee found that there was not defined a dose at which there was no testicular damage in animals. It was not possible to calculate what would be a safe way of using this substance. In the United States, DBCP was widely used and did cause testicular atrophy and sterility in the workers manufacturing the substance. I should like to say that the Pesticides Safety Precautions Scheme in the U.K. has been remarkably successful (M.A.F.F. 1979; Royal Commission on Environmental Pollution 1979; Perring & Mellanby 1977). We do not have deaths from normal use of pesticides in the U.K.

In the United States, several people are poisoned every year from pesticide use, mostly from parathion. In the U.K. there is no evidence of illness from normal use of pesticides, and only minor effects of accidental overexposure have been found.

This relative success does not absolve us from thinking about the problems of regulation. In particular we have the problem that when we consider social costs these are to some extent arbitrary in the sense that it is quite easy for one of the social groups involved to say suddenly, 'we will raise the costs of using this substance'. For the firms that manufacture it there are commercial restraints, but if trade unions say, 'we will not permit our members to work with this substance', then of course the cost–benefit equation has been altered very radically. If one is to have a system of regulation that serves the society and protects people, some kind of negotiation that brings together the people who are going to use the substances in a reliable way is needed.

Attitudes to the use of 2,4,5-T

The attitude of individuals to the same physical injury varies with the circumstance. Injury following an accidental event, such as being struck by lightning, is accepted far more readily than the same injury inflicted by a person, particularly if malice or carelessness can be attributed, or humiliation or powerlessness are felt.

Furthermore, if compensation can be claimed, and the sum depends on the extent of disability, then normal recovery may not take place and a life of embittered morbid obsession develops (see Hunter 1975, p. 1067). It does a major injury to a person to tell them falsely or carelessly that an event that was previously regarded as a natural and tragic misfortune, such as a cancer or the birth of a deformed child, was due to a man-made hazard, and due to negligence by employers or regulatory bodies. At the same time it is a fearful thing to be responsible for the release of a chemical substance that then injures people. How can such problems be resolved when the issue is treated as an area of political conflict?

2,4,5-Trichlorophenoxyacetic acid (2,4,5-T) is a selective herbicide. It has low toxicity to mammals in acute tests but causes abnormalities in newborn mice when the pregnant mouse is dosed early in the pregnancy. The effect is found even at low doses (M.A.F.F. 1981). No mutagenic or carcinogenic properties have ever been demonstrated and the teratogenic effects have never been found after exposure of the male to 2,4,5-T (only mutagens alter foetal development via the male). 2,4,5-T has been used in the U.K. for over 30 years because of its useful property of killing off broad-leaved scrub but leaving the grass to grow and cover the ground. No health problems have been noted in spite of contamination with tetrachlorodibenzodioxin (TCDD), which was probably present at around 5 mg/kg 2,4,5-T in the past, and is now less than 0.01 mg/kg. Recently there have been suggestions that the use of 2,4,5-T as a herbicide leads to injury in workers and their families.

It has been claimed that a number of miscarriages of pregnancy and birth defects in the children of agricultural workers were due to exposure to 2,4,5-T. On investigation, simple enquiries revealed that many of the men had not in fact used or been exposed to 2,4,5-T. An injury has been done to these families in the wrongful attribution of their problems. A different injury has been done if we find that 2,4,5-T has caused illness or birth defects in other cases.

Since around 50 % of all conceptions end in miscarriage, mostly in the very early days of pregnancy, and since around 2 % of all newborn children have clinically detectable abnormalities, it is not surprising that some families of agricultural workers suffer in this way, including those who have used 2,4,5-T. It is of great importance that proper epidemiology should be done to decide who is at special risk and why.

The type of regulatory systems that I have described are reasonably devised. Numerous variants of such systems exist, each with its own national rules and variations depending on whether the substance is classified as a drug, pesticide, food additive or industrial chemical. There is a gradual approximation of the rules set up by agencies in the U.S.A., E.E.C., O.E.C.D. and national authorities. Guidelines such as those issued on carcinogenicity, mutagenicity and testing of pesticides, food additives and industrial chemicals by U.K. government agencies have a considerable degree of coherence.

P.S.P.S. is fairly advanced in this field because it undertakes not only assessment of risk, but also stipulates what formulation may be used, on what crop, and how great an acreage may be treated. Similarly, the Committee on Safety of Medicines stipulates what claims may be made for new medicines. However, for medicines where it is possible to follow up what happens, there are adverse effects to every new drug. We must expect the same for other chemicals.

Origins of chemical toxicity

The most obvious cause of harm to man from chemicals is unexpected toxicity (table 7).

Testing procedures may not uncover an effect to which man is liable (Connell 1979). For instance, practolol was a useful drug against angina but produced fearsome side effects in fibrosis in the peritoneum and in other tissues. We still do not have a good test system whereby we could, in the future, prevent a drug with this kind of side effect from getting through the testing system.

Table 7. Origins of chemical injury

1. unexpected toxicity (practolol; MBK)
2. unexpected dose: management fault (Epping; Morocco; Michigan)
3. accepted risk: ? wrong person (parathion; melphalan; pertussis)
4. false perception of risk (cigarettes)
5. social loss from false overestimate of risk (chloroform; chromoglycate; 2,4,5-T ?)

Hexane, and in particular the 2-keto derivative, produces damage to peripheral nerves in people heavily exposed. Methylbutyl ketone (MBK) was used in glues and in some printing and in cotton dyeing processes. Hexane is present in all the petroleum fractions that have been in use for many years (Schaumberg & Spencer 1978). How is it that the toxic effects were not detected? Rats have to be exposed to hexane by inhalation for a long time before any serious injury develops and then it was quite difficult to pick up. Complex electrophysiological techniques and electronmicroscopy were used, so that it is not the kind of test that could be applied routinely to all substances. The neurotoxic effects were eventually discovered in an outbreak of nerve damage in a factory in the U.S.A. It is vital that we should have epidemiological surveillance and alertness to check that our toxicity testing is correct in scope and that the procedures used give the right answers.

The next major cause is unexpected exposure. That is, a substance was regarded as safe for one purpose and it was used for something else, or else was used in a way that led to much larger exposure doses. So for instance in Morocco, jet engine lubricating oil containing triorthocresylphosphate was sold as cooking oil, and many persons were permanently paralysed.

Failure of management and men to reduce exposure to known toxic substances is a common problem, both in working procedures and in accidents. Laboratory investigations of toxicity will not prevent mismanagement, but the development of toxicology as a subject in universities should be urged, to increase the awareness of the main factors in toxicity among all sections of the community (D.F.G. 1980; Lawless 1977).

A further source of toxic effects is when the risk and benefit are correct in the statistical sense but not for the individual. For instance, in the treatment of rapidly fatal ovarian cancer, the effective drug melphalan improves the chances of survival. But about 1 in 50 of those who survive for four or five years go on to develop leukaemia. So one can say before one starts that it is a risk worth taking. If we knew who would develop leukaemia we might use a different drug.

For whooping cough the risks and benefits are more subtle. There is a risk in vaccination. When whooping cough is a common disease, the risk of the disease is very much greater than the risk of vaccination. As the population becomes vaccinated, so the incidence of the disease falls and eventually the risk of vaccination becomes greater than the risk of the disease. However, if the population is not vaccinated the incidence goes up. Of course, the lowest risk option is if everyone *else* is vaccinated (Ambrosch & Wiedermann 1979).

For asbestos, fire doors save lives. But it may be that a different group is at risk from those that benefit. That is a fundamental social and political problem and I think that the job of the scientist is to make clear what are the options, what are the consequences of different actions, and in particular to point out at times of conflict that there may be alternative substances that can be used, whereby the conflict can be resolved.

Safety margins

For food additives it is common to have a 'safety margin' whereby the 'acceptable daily intake' is set to be one-hundredth of the highest dose causing no adverse effect in the most sensitive species, the factor of 100 supposedly made up of 10 for interspecies variation and 10 for inter-individual variation in dose and in response.

For many foods such as modified starches or novel proteins, fed at up to 5% of the diet, no such margin can be possible, and for contaminants such as lead, cadmium or nitrate we are very much closer to the toxic levels than is desirable.

For medicines, desperate diseases need desperate remedies, and both surgery and medical treatment of serious illness may work at levels where death is possible and serious side-effects common. For less serious conditions, such as home remedies for headaches, serious injury may be found at 10–50 times normal dosage.

For pesticides the most common problem is that while proper use should lead to minimal exposures of public and workers, several hundred times below the toxic levels, the real question is how much weight should be given to the possibility of misuse, accidental exposure, or use in suicide or homicide. For one pesticide, formulation with an evil-smelling 'stenching agent' has reduced the homicidal risk, while for another highly toxic active compound, with an $\mathrm{l.d.}_{50}$ in the region of 1 mg/kg, the risk was made negligible by dilute formulation into salt pellets. This preparation could produce human exposure only if many of the salt pellets were dissolved. So the question of 'safety margins' becomes one that can be answered only in the context of who will use the substance, for what purpose, to what benefit, with what precautions, and with what information and consent.

False perceptions

Another source of injury comes from the false perception of risk. An obvious example is young people riding motor cycles, who almost certainly do not have a balanced view of this risk, nor those of smoking or alcohol. Those are personal choices. Much more serious is the problem if the manager in charge of a chemical plant has a false perception of the risk. So it is most important that people should become informed. Then there is another risk, which is the social risk of costs of diversion of resources into the wrong places. For instance, feeding phenobarbitone produces tumours in the mouse liver. We have a long history of human use of phenobarbitone without any liver tumour developing and the result is that we continue to use this valuable drug, but it could very well have been that experts taking seriously the wrong model system would think that we must ban phenobarbitone (White *et al.* 1979; Costello & Snider 1980). That would be an error. In the end the necessary social costs are liable to be reflected in reduced real resources for people, which should be used to improve health.

Another example is in the chlorination of water supplies. The process is most important in reducing bacterial water-borne disease. Chlorination leaves a very small residue of chloroform and other chlorinated carbon compounds in the water. Chloroform in large doses causes cell injury and tumour formation in the liver and kidney and other tissues in mice. It is likely that some cell injury is necessary for these effects to take place, so that if the dose is low enough, no cell injury takes place and there will be no tumour formation. It is also likely that the mouse is the wrong kind of animal. If the use of chlorine as a water-sterilizing agent is banned, the developed industrial countries can achieve the same end by other more expensive means. Less-developed countries may follow the lead in banning chlorination without having effective alternatives. This could be a greater disaster than the one that overtook Sri Lanka when the use of DDT was banned and malaria returned.

We started off by saying that when society has large resources it is able to create conditions for better health. Now we see a conflict between more resources like better plastic drain pipes, with drier, warmer houses, and on the other hand, chemical risks, especially for workers. We started off with a simple view of our testing procedures; if it harms a mouse it probably harms us. As dose and species variation become clear it becomes necessary to use pharmacokinetic and metabolic studies to correct for variation between species and to calculate the effects on man of low dose exposure. Now we become more sophisticated and use other model systems, say bacterial mutations or isolated tissues, brain slices or synaptosomes from the central nervous system: the concept of harm to bacteria or synaptosomes is no longer valid.

The test system now requires validation. We say that substances A, B and C harm man and they do X and Y to our model test system. Other substances which also do X and Y to the test system probably harm man, but that is not necessarily true, because many substances that do not harm man may also do X and Y. So each test system requires careful validation. That means that we need epidemiology to do toxicology. We need to have studies of the effects of substances on man. We constantly have to verify the correctness of our predictions for new substances, and to validate the use of our model systems. We finish up with a triple system of looking at toxic chemicals. We have a set of toxicity tests and measurements so that we can measure toxic effects in animals and model systems. We do pharmacokinetic metabolic studies in animals and in man to find the true dose to the target organ and how this varies with the use of the substance. Clearly toxicology cannot be properly done in isolation: it makes little sense to have a toxicological assessment of a substance if you do not know what it is going to be used for. Thirdly, we need epidemiology to verify our predictions and to keep our test systems clean and in order.

With these we can give reliable estimates of the risks involved in the use of chemicals. The evaluation process, whether the risk is justified, is a matter for the social decision process.

Toxicology has to develop as a science, putting order and meaning into our observations of biological systems exposed to chemicals. To do this it has to draw on many basic subjects from chemistry to statistics and pathology. But toxicology is also concerned with harm, which is a human and social dimension. Toxicologists have the dilemma of choice of emphasis between bringing order and knowledge to their developing subject, or else using what is known in practical terms.

In toxicology it becomes particularly clear that theory and practice have to develop together if the subject is to develop as an exciting and useful branch of science.

References (McLean)

Ambrosch, F. & Wiedermann, G. 1979 *Dev. biol. Stand.* **43**, 75–83; 85–90.

Chenery, R. J. & McLean, A. E. M. 1980 *Biochem. Pharmac.* **29**, 271–276.

Connell, P. H. 1979 In *Side effects of drugs annual*, vol. 3 (ed. M. N. G. Dukes), p. 3. Amsterdam: Excerpta Medica.

Costello, H. D. & Snider, D. E. 1980 *Am. J. Epidemiol.* **111**, 67–74.

D.F.G. 1980 *Memorandum toxicology.* Deutsche Forschungsgemeinshaft Boldt Verlag Boppard

D.H.S.S. 1980 In *Inequalities in health*, p. 75. London: Department of Health and Social Security.

Doll, R. 1979 *Proc. R. Soc. Lond.* B **205**, 47–61.

E.P.A. 1979 *Trifluralin position document.* Special Pesticide Review Division, Office of Pesticides Programs, Environmental Protection Agency, Washington.

Garner, R. C. & McLean, A. E. M. 1969 *Biochem. Pharmac.* **18**, 645–650.

Gehring, P. J., Watanabe, P. G. & Park, C. N. 1979 *Toxicol. appl. Pharmac.* **49**, 15–21.

Heavey, C. D., Ury, H., Sieglaub, A., Ho, K. P., Salomon, H. & Cella, R. L. 1980 *J. natn. Cancer Inst.* **64**, 1295–1299.

Hoel, D. 1980 *Fedn Proc. Fedn Am. Socs exp. Biol.* **39**, 73–75.

Hunter, D. 1975 *The diseases of occupations*, 5th edn. London: E.U.P.

Korte, F. 1977 In *The evaluation of toxicological data for the protection of public health* (ed. W. J. Hunter & J. G. P. M. Smeets), pp. 235–246. Oxford: Pergamon.

Lawless, E. W. 1977 *Technology and social shock.* New Brunswick: Rutgers University Press.

McLean, A. E. M. 1979 *Proc. R. Soc. Lond.* B **205**, 179–197.

McLean, A. E. M. & Day, P. A. 1975 *Biochem. Pharmac.* **24**, 37–42.

M.A.F.F. 1979 *Pesticide Safety Precautions Scheme.* London: Ministry of Agriculture, Fisheries and Food.

M.A.F.F. 1981 *Further review of 2,4,5-T.* London: Ministry of Agriculture, Fisheries and Food.

O.P.C.S. 1978 *Occupational mortality, 1970–72*, Decennial supplement, p. 109. London, H.M.S.O.

O.P.C.S. 1980 *General household survey, 1978*, p. 133. London: H.M.S.O.

Perring, F. H. & Mellanby, K. (eds) 1977 *Ecological effects of pesticides.* London: Academic Press/Linnean Society.

Peto, R. 1980 *Nature, Lond.* **284**, 297–300.

Registrar General 1967 *Statistical review of England and Wales, 1966*, part 1: Tables medical. London: H.M.S.O.

Royal Commission on Environmental Pollution 1979 *7th Report: Agriculture and pollution.* H.M.S.O. London.

Schaumberg, H. H. & Spencer, P. S. 1978 *Science, N.Y.* **199**, 199–200.

Spear, R. C., Popendorf, W. J., Spencer, W. F. & Milby, T. H. 1977 *J. occup. Med.* **19**, 411–414.

Stanley, P. & Bunyan, P. J. 1979 *Proc. R. Soc. Lond.* B **205**, 31–45.

White, S., McLean, A. E. M. & Howland, C. 1979 *Lancet* ii, 458–460.

Williams, M. H. C. 1958 In *Cancer* (ed. R. W. Raven), pp. 337–380. London: Butterworths.

Zbinden, G. 1980 *Swiss Chem.* **2**, 23–28.

Proc. R. Soc. Lond. A **376**, 65–78 (1981)
Printed in Great Britain

Mortality statistics and the assessment of risk

By A. J. Fox

Department of Mathematics, The City University, Northampton Square, London EC1V 0HB, U.K.

For nearly 150 years, mortality statistics have provided the base that epidemiologists have used to quantify and evaluate risk. In early years these derived directly from information recorded on death certificates, which enabled medical statisticians to calculate periodically death rates for various causes of death by sex, age, area of residence and occupation. With time, interest moved from the immediate effects of an occupation to more delayed effects. Mainly as a response to this interest, the last 30 years have seen developments in national record systems that allow epidemiologists to mount prospective mortality studies capable of answering detailed questions about the nature of occupational risks and describing dose–response relations. This paper outlines these developments, thereby providing a background for the following paper, which describes the use of these methods in occupational studies of dose–response relations.

Background

Death rates are of intrinsic interest in the quantification of risks, primarily because death is absolute and because for a number of years data on all deaths have been systematically recorded. However, deaths and death rates can never completely describe the risk process. The contribution of mortality statistics to this topic is limited first by the extent to which mortality describes the 'harm' resulting from particular 'hazards' and second by the scope of the data collection systems. While the main focus of this paper concentrates on the data collection systems, describing recent developments, a brief description of the risk process should indicate the limited contribution of mortality data to an overall perspective on risk.

The risk process

The terms 'hazard', 'risk' and 'harm' are used here to describe different facets of the risk process. 'Hazard' and 'harm' describe 'exposure' and 'effect', respectively. 'Risk' is a matrix of probabilities linking various elements of exposure to those of effect. In this model, 'hazard' and 'harm' have a number of dimensions, many of which correspond to the dimensions already identified at this meeting as being important to the assessment and perception of risk. These dimensions include (*a*) for hazard: cause of hazard, size of group affected, who carries risk, and experience of risk; and (*b*) for harm: outcome, timing of outcome, and duration of outcome.

Many of the hazards that man faced in his early struggle for survival may be described as natural hazards in that they reflect the hazards of his environment; from animals, from infections and viruses, or from floods and earthquakes. However, even in these early times, groups of men would fight each other, and to that extent some hazards were man-made. With the development of modern technological society the hazards that man faces have become increasingly complex. The hazards associated with transportation and with industrial production are often considered to be man-made. Some are created by the individual (for example, in the way he drives his car or operates his machine); others reflect more on the demands of society (for example, in the degree of protection against hazards that people are prepared to pay for). The suggestion that one can protect against hazards introduces the idea that some hazards are avoidable whereas others may be unavoidable. The possibility of the Thames flooding is an example of an avoidable natural hazard.

The scale of hazards in which we are involved has been changing. Societal concern appears recently to have moved from hazards that affect individuals or small groups of people to hazards, mainly man-made, that can lead to large numbers of people dying. For many people, the reawakening of the nuclear debate emphasizes the risk of man even destroying whole societies.

Two further dimensions of hazard have also been found important in discussion of risk perception. The first concerns the difference between hazards to self and hazards to other people; mothers, for example, will often be more concerned about hazards to their children than about hazards that they themselves face. Alternatively, people living near to a chemical plant complex or working with hazardous materials may have different perceptions of the hazards than those at minimal risk of being affected.

The second dimension concerns the experiences that the individual has of the hazard in question. Most people are, for example, aware of the hazards associated with crossing roads or driving cars, because these are well publicized and their harmful effects are widespread enough for people to come into contact with them. Some hazards, such as those associated with transportation by rail or air, are known about but accidents are nevertheless rare and are therefore of less concern to the individual. However, public interest has increased in hazards of which society has no, or limited, experience. Examples include that of a major disaster at a complex of chemical plants, such as at Flixborough, or of a major leak, or explosion, at a nuclear power station. In such examples, we not only have little idea about the probability of an adverse event occurring but also we can only speculate as to the physical and social consequences.

As already pointed out, risk analysis has, in the past, been mainly concerned with the risk of death, primarily because data on causes of death were readily available. However, much recent interest has developed in measuring the risk of other outcomes, such as human injury or disability, death or injury to animals and livestock, damage to the environment and amenity or simply financial loss.

The effects in which we are mainly interested are not always observed at the time that the relevant exposure takes place. For accidental deaths, when the time between cause and effect is short, it is often relatively simple to identify and describe the causes of the accident. However, for deaths from chronic diseases, when the 'latent interval' may be several years, this process is not straightforward. First, a number of alternative factors may contribute to the same disease and, secondly, the timing may depend on disease mechanisms that are not clearly understood.

Not only may the effects measured be of different kinds, and occur at different times, but they may also be of different levels of duration. Clearly, death is irreversible, but with some effects the disability or damage to the environment may be cured or repaired. For a number of effects, however, the disease process, once started, may be irreversible and may even lead to progressively deteriorating health after the event.

I apologize if this brief description of the problem area goes over ground covered by other speakers, but I believe that it is essential for a balanced perspective that the dimensions of 'hazard' and 'harm' be emphasized before numerical data are presented. There may otherwise be a tendency to ignore these dimensions and to consider the data presented as, in some sense, absolute. There should now be no misunderstanding of the relatively limited way in which the available statistics describe many of the processes with which we are concerned. The remainder of this paper should, however, indicate how, within the limited scope of mortality, methods have been developed that enable epidemiologists to describe various aspects of the risk process.

Cross-sectional unlinked mortality rates

John Gaunt and William Petty first used death certificates to calculate death rates as long ago as 1662. Since 1837, death records for England and Wales have been systematically brought together by the Registrar General, who publishes annual statistics giving the numbers of deaths by sex, age and cause of death. Farr, the first compiler of the abstracts in 1839, recognized the importance of relating the numbers of deaths in each cell to the population alive at the time. Using data derived primarily from decennial census (updated to allow for intervening births, deaths and migrations) Farr calculated sex, age and cause of death specific death rates.

At regular intervals the Registrar General extends the scope of these tables to incorporate analyses by marital status, by area of usual residence and by occupation. The death record and the corresponding census record provide only a cross-sectional description of the characteristics of people who die and those who are alive. Also, because records for the same individual are not linked together, a number of biases between numerator and denominator can affect the statistics.

The limitations of these data are not confined to those deriving from numerator–denominator biases. The latest review from O.P.C.S., published in 1978, points to three other major weaknesses that also affect more sophisticated studies using methods of record linkage. These three weaknesses, all recognized in the last century, are: (*a*) difficulty in differentiating between intrinsic and extrinsic factors associated with occupation; (*b*) selection for work and survival in the occupation; and (*c*) long-term, as opposed to immediate, effects.

Arlidge separated the conditions and circumstances of labour into incidental or apparently essential (intrinsic) factors and accidental or non-essential (extrinsic) factors. The former include the type and circumstances of the work performed and the materials and machinery used. The latter are composed of the personal and social characteristics of people following particular occupations. For example, the lung cancer mortality of men following a particular occupation may be high either because these men are exposed to carcinogenic chemicals at work, because they smoke excessively or for a combination of these two reasons.

Ogle pointed to the importance of selection processes determining the health of people who follow particular occupational paths. These processes affect not only the choice of job made by the individual, but also his acceptability to an employer. Similarly, as his health deteriorates, so may his chances of continued employment. Clearly these processes operate at different levels according to the type of work and the circumstances in which it is to be performed.

Interest in long-term as opposed to immediate hazards has increased in recent years. Nevertheless, concern about the damage to miners' lungs caused by high levels of dust exposure and cancer of the scrotum among chimney sweeps are just two examples of early interest in the delayed effects of work-related exposure.

With these limitations in mind, these data can be looked at in two ways. The first is by concentrating on each individual occupation to see whether or not there is evidence of any peculiarities in the mortality pattern observed. Is the mortality from any individual cause of death or group of causes high? There are many examples where this type of analysis confirms widely recognized occupation-related effects.

Alternatively, particular causes of death can be considered to discover if there is a common characteristic in those occupations that have high mortality from those particular causes. Analyses in the latest review point to relations between accidental falls and work at height; between transport accidents and work for example on boats, trains and planes; between suicide and access to drugs; between cirrhosis and alcohol consumption; and between dusty work and cancer of the stomach.

The need for more meaningful measures of exposure

How, then, do these cross-sectional, unlinked analyses of mortality by occupation help in providing measures of risk? Returning to our original simple model relating hazard and risk to harm, the mortality rates that these supplements provide indicate the proportion of people following a particular occupation in a particular year who die in that year. By a comparison with national death rates, an attempt is made to attribute excess mortality from particular causes to the occupation of interest. However, the approach gives no guidance as to any time relations between exposure and effect, nor, more importantly, does it help point to relevant exposures to which people following each occupation may be subject. Without these components its use is restricted to the generation of hypotheses about the existence of occupation-related hazards; its contribution to the study of these relations is minimal.

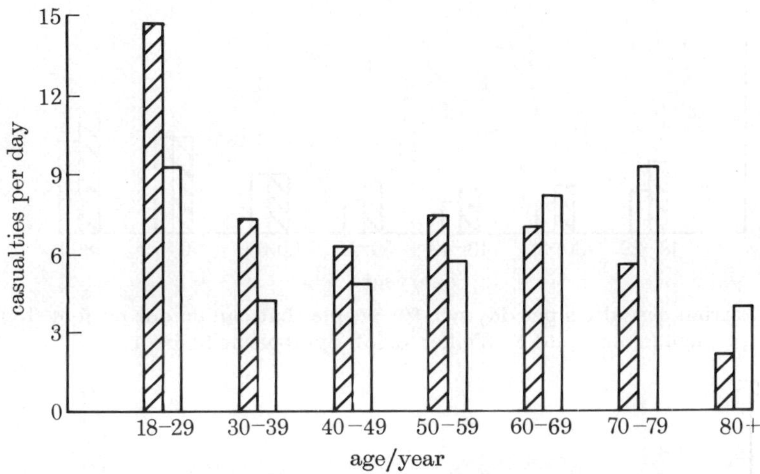

FIGURE 1. Pedestrian casualties per day, by sex and age (from Todd & Walker 1980). Hatched bars, females; open bars, males.

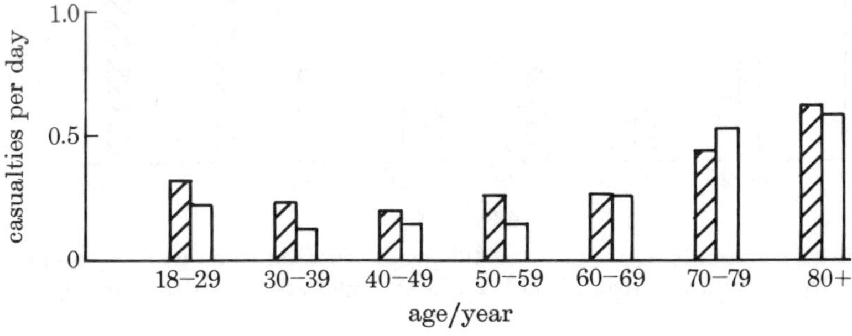

FIGURE 2. Pedestrian casualties per 10^5 population per day (from Todd & Walker 1980). Symbols as figure 1.

A recent survey carried out by O.P.C.S. for the Transport and Road Research Laboratory illustrates the type of data that must be collected to enable the particular hazard–harm relation to be pursued. The survey, entitled 'People as pedestrians', was conducted to ascertain to what extent people go out and about on foot. Figures from the survey were compared with pedestrian casualty figures to estimate casualty rates by sex and age and, subsequently, by day of the week and by time of day. Figures 1–4 illustrate how, by refining the measure of who is exposed, a different picture of the relative risks, particularly in relation to age, is obtained. The argument is further extended in table 1, which shows the effect not just of identifying who is exposed, but also of incorporating a measure of the duration and type of exposure. The last three rows of table 1 are further supported by figure 5, which highlights the effect of considering all pedestrian casualties without separating those occurring when the pedestrian was crossing a road from those where he or she was not crossing a road.

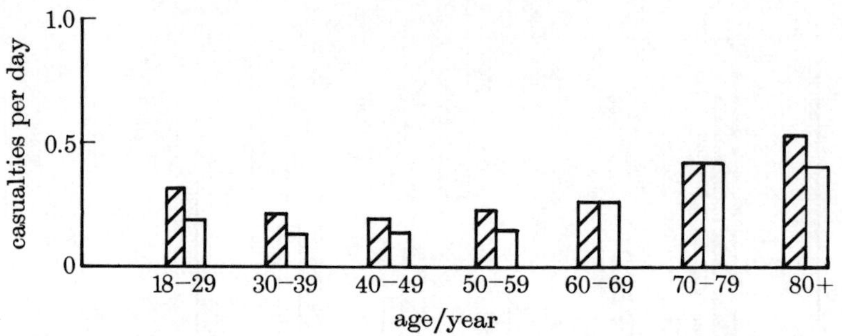

FIGURE 3. Pedestrian casualties per day per 10^5 people that can go out on foot, by sex and age (from Todd & Walker 1980). Symbols as figure 1.

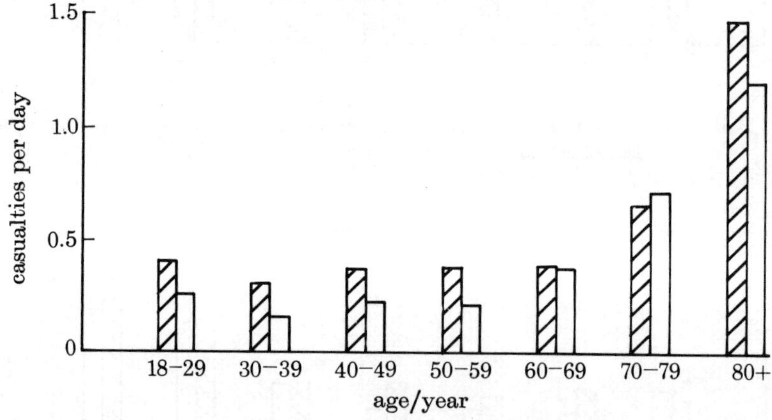

FIGURE 4. Pedestrian casualties per day per 10^5 people that do some walking on an average day, by sex and age (from Todd & Walker 1980). Symbols as figure 1.

PROSPECTIVE MORTALITY STUDIES

As long as mortality statistics were derived without linking the death record to an exposure record, little progress could be made in the range of mortality statistics available. Without record linkage, analyses are restricted to the characteristics recorded at the time of death. In the early 1950s several epidemiologists, in particular R. A. M. Case and Sir Richard Doll, realized that, to assess risks attaching to particular exposures or work situations, it would be necessary to identify groups of men whose exposure was properly documented and to follow those groups forwards in time. They mounted the now well known prospective mortality studies of rubber and chemical workers, of asbestos workers and of doctors; the latter by looking at their mortality in relation to cigarette smoking. In such early studies, elaborate procedures were developed to monitor the cohort for deaths and to confirm that other individuals were alive at the end of the follow-up period. Thanks primarily to facilities developed by O.P.C.S., using the National Health Service

TABLE 1. WHAT DO DIFFERENT MEASURES OF RISK TELL US ABOUT THE RELATIVE RISKS BETWEEN THE OLD AND THE YOUNG?

(From Todd & Walker (1980).)

measure of pedestrian casualty risks (per day)	ratio of risks at ages 80 years and over to those at ages 18–29 years
mean pedestrial casualties	0.3
per 10^5 population	1.7
per 10^5 people who can go out on foot	2.4
per 10^5 people who do some walking on an average day	4.1
per 10^8 km walked	7.1
per 10^8 h spent walking	3.7
per 10^8 roads crossed	6.9
per 10^8 road sections crossed	6.8
per 10^8 interviewer road crossing paces	6·9

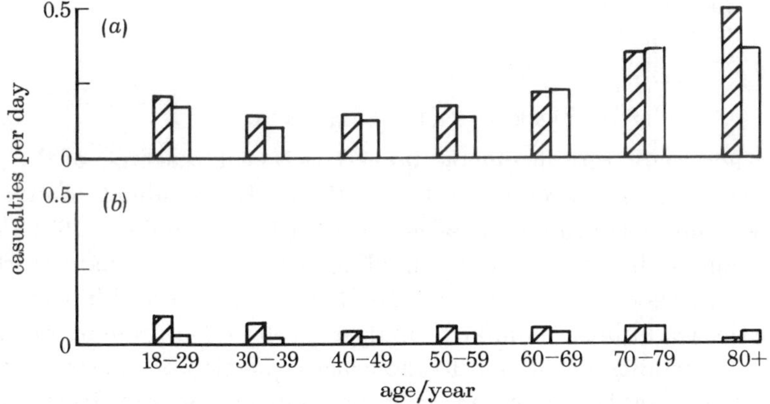

FIGURE 5. Comparison of pedestrian casualty rates per day per 10^5 population, according to whether pedestrian was crossing the road (*a*) or not (*b*) (from Todd & Walker 1980). Symbols as figure 1.

Central Register at Southport, it is now relatively easy to monitor the mortality of any identified cohort. Also, these record systems enable epidemiologists to mount historic studies following the mortality of people exposed a number of years in the past.

The approach is now widely used: O.P.C.S. is assisting with more than 100 separate studies, and as a result much research has now been devoted to studies of detailed relationships between exposure and effect. In the next paper, Acheson describes the use of these methods by epidemiologists. My concluding section will indicate how such methods are now contributing to improvements in national statistics, giving, as an illustration, an analysis showing how these can be used to shed light on factors determining the risk of death associated with riding a motor cycle.

Table 2. Overall mortality in 1971–5 of men employed in 1971 by means of transport to work in 1971

transport to work	deaths		
	observed	expected	s.m.r.
car	1670	1754.7	95
public transport	1366	1287.1	106
foot and bicycle	1117	1127.1	99
motor cycle	87	73.1	119

Table 3. Mortality from road accidents in 1971–5 for men employed in 1971 by means of transport to work in 1971

transport to work	road accident deaths		
	observed	expected	s.m.r.
car	44	47.5	83
public transport	22	30.9	71
foot and bicycle	28	23.7	118
motor cycle	12	3.0	400

Routine-linked statistics

For the past few years, routine demographic data on a 1 % sample of the population have been brought together in the O.P.C.S. Longitudinal Study. This source includes a link between deaths after the 1971 Census and the 1971 Census record for the same individual, thereby enabling O.P.C.S. to calculate mortality rates by the full range of census questions. To illustrate how record linkage in the study adds to the traditional 'unlinked' analyses, tables 2–5 present the results of an analysis of mortality in relation to the census question on 'means of transport to work'. These tables are based on very small numbers of deaths, and I should emphasize that they are only given to illustrate an approach.

Tables 2 and 3 give the numbers of deaths in 1971–5 to men who in 1971 were employed, according to how they travelled to work in 1971; table 2 gives deaths from all causes, table 3 gives deaths from road accidents. To account for differences in age structure between the various groups, expected deaths are calculated on the basis of the age structure of the group and standard death rates for all employed men.

As might be expected, there was little difference between the major groups in overall numbers; there is a suggestion that men who went to work by car had a slightly lower mortality than those who went to work by public transport. By taking account of the number of cars to which the household had access, this can be shown to be a social class effect. The overall mortality of men who in 1971 travelled to work on a motor bike was raised by some 20% (the 95% confidence limit was from 98 to 140). A clear difference is, however, seen in table 3, which focuses on road accident deaths. The s.m.r. for motor cyclists is now some four times that for all employed men (95% confidence limit from twice to seven times).

TABLE 4. ROAD ACCIDENT DEATHS IN 1971–5 FOR MEN WHO IN 1971 WERE EMPLOYED AND TRAVELLED TO WORK ON A MOTOR CYCLE, BY AGE

(O.P.C.S. Longitudinal Study.)

age/year ...	15–19	20–24	25–44	45–64
	road accident deaths			
motor cyclists	5	2	2	2
all employed men	10	14	43	33
	rates (per 1000 person-years at risk)			
motor cyclists	3.03	0.69	0.43	0.41
all employed men	0.50	0.23	0.15	0.13

TABLE 5. ROAD ACCIDENT DEATHS IN 1971–5 FOR MEN WHO IN 1971 WERE EMPLOYED AND TRAVELLED TO WORK ON A MOTOR CYCLE, BY AGE AND YEAR OF DEATH

age/year ...	15–19	20 and over
	road accident deaths	
motor cyclists		
1971	4	3
1972–5	1	4
all employed men		
1971	4	17
1972–5	6	85
	rates (per 1000 person-years at risk)	
motor cyclists		
1971	8.16	1.73
1972–5	0.86	0.34
all employed men		
1971	0·65	0·18
1972–5	0·44	0·16

This example shows the prospective study approach in which the epidemiologist measures relative risk by identifying a group of people at risk (in this case men who in 1971 travelled to work by motor cycle), and comparing their mortality forwards in time with that of a standard group. The analysis may then subdivide the people at risk by factors that were recorded when the population was identified and which may be related to increased exposure or risk.

Although our example is based on very few deaths, table 4 breaks these road accident deaths down by age to show how the main excess is concentrated among men aged 15–19 years. It is interesting to note from this table that of the 10 road accident deaths in 1971–5 to men aged 15–19 years, five were to men who in 1971 travelled to work by motor cycle. National statistics confirm that approximately 60 % of the 850 road accident deaths per year to men aged 15–19 years are to men riding motor cycles at the time of the accident. The findings are even more striking when deaths in 1971 are separated from those in 1972–5. Table 5 shows that among the 15–19 year old men the death rates for 1971 were nearly ten times those in 1972–5. In this 1 % sample, all road accident deaths in 1971 at age 15–19 years were to men who travelled to work by motor cycle.

The difference in risk between 1971 and 1972–5 may reflect men giving up riding a motor cycle after a relatively short exposure. It may be that the men, who at 16 years were only allowed to ride a motor cycle, transferred to driving a car at age 17 years. However, within the limitations of the small numbers, the data in table 5 would suggest that the high risk in 1971 probably reflects the risk associated with inexperience. The same pattern of decline in risk between 1971 and 1972–5 is seen for men aged 20–24 years, who would not be affected by the minimum age for driving a car. Also if there was a tendency for men to stop riding a motor cycle after short exposure we would expect to see a similar decline between 1972 and 1973, etc.; but this is not evident when the data are further subdivided. This suggestion that inexperience is a major factor in the level of risk for motor cyclists is further supported by a recent survey by the Transport and Road Research Laboratory. Their review of 235 motor cycle accidents indicated that 85 (36 %) involved people who had been riding for less than 6 months; 59 (25 %) had been riding for less than 3 months. Although these percentages are high, the T.R.R.L. do not have denominators with which to compare these figures.

It would be possible, by using census records, to extend this study to include a larger sample of motor cyclists. However, although this may confirm the differences noted here, it would still not shed light on the explanation of these differences. For that we should need to mount a prospective study based on more detailed exposure measures, including, for example, how experienced the rider was and the average mileage per year. However, because death and census records were linked in our example, we have been able to use this routine record system to go a little further than routine unlinked statistics and to suggest how risks associated with riding a motor cycle may vary among cyclists.

Summary

This paper emphasizes the importance of being aware of and understanding all the dimensions of the risk process of interest. Mortality data are routinely published, but these often do not enable researchers to derive the measures that they desire. Recently, record linkage methods have made better use of situations in which exposure has been properly documented. At the same time, the introduction of these methods into national statistical systems should mean that these should also now be expected to contribute more relevant and reliable data. However, despite these improvements in the measures that we can obtain, it should be recognized that simple probabilistic measures of risk will rarely adequately describe the risk process – that is, the full relation between hazard and harm – and that the diverse dimensions of hazard and harm make risk comparison extremely complex.

BIBLIOGRAPHY (FOX)

Case, R. A. M., Hosker, M. E., McDonald, D. B. & Pearson, J. T. 1954 *Br. J. ind. Med.* **11**, 75–104.

Cullen, M. J. 1975 *The statistical movement in early Victorian Britain.* New York: Harvester Press.

Doll, R. 1955 *Br. J. ind. Med.* **12**, 81–86.

Office of Population Censuses and Surveys 1973 *Cohort studies: new developments.* (*Studies in medical and population subjects*, no. 25.) London: H.M.S.O.

Office of Population Censuses and Surveys 1978 *Occupational mortality*, 1970–72. (*Decennial supplement* DS no. 1.) London: H.M.S.O.

Todd, J. E. & Walker, A. 1980 *People as pedestrians.* London: H.M.S.O.

Discussion

M. J. GARDNER (*M.R.C. Environmental Epidemiology Unit, Southampton, U.K.*). Professor Fox showed that among persons who were motor cyclists at the time of the 1971 Census, the road accident death rate decreased during 1972–5 from that in the later months of 1971. He attributed this decline to the benefit of the learning experience of continued motor cycling. An alternative explanation of these data would seem to be that a fraction, possibly large, of the enumerated motor cyclists had ceased to ride their bikes in the following years and hence were removed from exposure to risk – except to transfer to the much lower risks of travellers by car, public transport or pedestrians.

A. J. FOX. I do apologize for not having mentioned this possible explanation at the meeting. The printed version of my paper acknowledges the possibility of a marked drop in the numbers continuing to ride motor cycles within a few months of the census. While it is likely that 16 year olds convert to cars at the age of 17 years, it seems less likely that older men will also give up at the same rate. However, the reduction in risk between 1971 and 1972–5 was also noted at ages 20–24 and 25 and over. The explanation that early increases in experience are associated with greater reductions in risk than later increases is therefore more plausible an explanation. This is not to deny that giving up does not contribute to this decline. For this reason the paper suggests that a prospective study of new motor cyclists, recording their experience and relating this to mortality, be mounted to quantify these two factors.

F. Lees (*Department of Chemical Engineering, Loughborough University of Technology, U.K.*). This is a comment rather than a question. In his presentation Professor Fox has described the use of death certificates and census returns as a data base for epidemiological studies of a wide variety of hazards, and has emphasized the generation of hypotheses as a preliminary to more detailed data collection and study. My comment concerns the specific hazard of long-term, low-concentration exposure to toxic chemicals in the working environment in the chemical industry. This is a hazard that must be taken very seriously. Epidemiology has a vital role to play in identifying such a hazard. Unfortunately, however, a toxic effect may be slow to develop. For such a hazard it seems unsatisfactory to rely on a two-stage process in which the first stage is hypothesis generation and the second a full epidemiological study, and in which each stage is preceded by collection of data over a period measured in decades. It is very desirable, therefore, to collect from the beginning information that constitutes a data base capable of substaining a reasonably detailed epidemiological study. Works medical records should provide the elements of such a data base. My point is that epidemiologists have a crucial role to play in ensuring that the quality of the data in the data bases available for the study of long-term toxic hazards matches the importance of these hazards within the usual constraints of what is reasonably practicable.

A. J. Fox. In order not to overlap with the next speakers, my paper concentrated on the routine data that are collected by central government. As can be seen, the printed version of my paper describes facilities developed by O.P.C.S. that assist epidemiologists to use industrial records as the basis for more detailed studies. O.P.C.S. collaborates closely with researchers concerned with studying the effects of exposure to a wide range of substances; over 100 such studies are currently being undertaken. The Occupational Mortality Statistics, which O.P.C.S. produces every 10 years, are not directed to the situation where a hazard is suspected and where there are grounds for conducting a proper study. They are produced to provide a perspective, comparing *all* occupations, and, at the same time, to provide clues of as yet unsuspected hazards. We do not have the epidemiological resources – neither for follow-up nor for analysis – to insist that each industry and each sector of industry be the subject of continuous epidemiological investigations. For this reason, we need a multi-stage approach.

C. Green (*School of Architecture, Dundee, U.K.*). Death certificates may not be a wholly reliable data source as a basis for assessing risk. For example, the Institute of Sports Medicine was recently reported as having stated that there were four rock-climbing deaths in the U.K. in 1978. The British Mountain Rescue Committee reports 35 deaths in the same year. I suspect that I.S.M.'s figures are based upon death certificates and that, as a source, these under-reported rock-climbing deaths, not all deaths by, say, fractured skulls being recorded as having been sustained in a fall from a rock face. Equally the B.M.R.C.'s figures may be inflated by the inclusion of some hill walking deaths. Has Professor Fox any general comments upon the degree and sources of unreliability in death certificates?

A. J. Fox. O.P.C.S. analyses accidental deaths by the external cause of death, by the nature of the injury and by the place of the accident, based on information as recorded on death certificates. My paper tries to emphasize the limitations of analysing death certificates independently from the population exposed, and the limitation pointed to here is yet another example. However, by bringing the deaths together with the population at risk in prospective studies, many of the limitations to which this question alludes can be overcome.

If, for example, I were following a cohort of men known to be rock-climbers I would record an excess of deaths from falls, only a few of which would be recorded as rock-climbing deaths. In trying to explain this supposed excess of deaths from 'falls, other than in rock-climbing', I would be expected to investigate the possibility of under-reporting of rock-climbing deaths. My total risk from falls would not, however, be affected by the bias.

In the next paper, Professor Acheson describes how prospective studies have been used to study the relation between lung cancer and asbestos exposure. Few of the death certificates record 'lung cancer due to asbestos exposure'. Nevertheless, they provide an adequate basis for his study. In isolation, therefore, death certificates are of limited use in the quantification of risk, but when used in conjunction with an exposed population they provide the most powerful weapon in the epidemiologist's armoury.

D. Andrews (*Heathlands, Cawston, Rugby, U.K.*). Mr Fox has drawn attention to the improvement in information value of studies of subjects as individuals over studies of subjects in totality. A recent investigation into the safety records of Grand Prix drivers provides an excellent illustration of this, and incidentally provides information that is relevant to issues that will be discussed later.

In 1975, the Jim Clark Foundation published the results of a survey, based on eye-witness accounts, aimed at identifying the causes of the 224 recorded accidents that had occurred in practice and race sessions in the seven seasons 1966–72. Almost exactly half of these accidents were found to have been due to driver error.

A cursory inspection of these records revealed that the distribution of driver error accidents was quite different from the distribution of the accidents over which the drivers had had no control. I therefore divided the drivers into 10 groups of about 200 driver appearances each, ranging from 0 to 9 driver appearances each for 77 drivers at one end, to a total of 280 driver appearances for 3 drivers at the other. It was discovered that the fractions of driver error accidents experienced by these groups fell near a curve, from which it was possible to deduce that the mean propensity to incur driver error accidents fell by two orders of magnitude over the first 70 to 100 driver appearances for each driver.

Equally interesting, considering only those 31 drivers who had made 22 driver appearances or more, and for whom there was therefore a mean propensity to have incurred $1\frac{1}{2}$ driver accidents or more, it was found that the normalized distribution was triangular. At one extreme, 8 drivers had incurred no driver error accidents, while 1 had incurred nearly three times as many driver error accidents as might have been expected. By contrast, the distribution of accidents over which the drivers had had no control was uniform, and the normalized distribution had a standard deviation of 0.65.

It became possible to discover these trends only because the record was available for each individual driver. Any conclusion that might have been drawn from a global statistic about the tendency of drivers to incur driver error accidents would have been misleading.

Proc. R. Soc. Lond. A **376**, 79–85 (1981)
Printed in Great Britain

Dose–response relations from epidemiological studies

By E. D. Acheson and M. J. Gardner

M.R.C. Environmental Epidemiology Unit, University of Southampton, Southampton General Hospital, Southampton SO9 4XY, U.K.

The paper discusses dose–response relations by using information about carcinogens in man as examples. Setting aside tobacco, alcoholic drinks and certain drugs, there are only two factors for which quantitative data about personal exposure in industry to incidence of cancer are available over a range wide enough to make it possible to study the shape of the relation: chrysotile asbestos and ionizing radiation. In both cases the relation is considered to be linear. As far as chrysotile is concerned, extrapolation based on the linear model suggests that the risks of lung cancer associated with exposure to the levels of asbestos dust found in buildings not under construction or repair and in ambient urban air are trivial. Contrary to popular opinion, the risk of mesothelioma increases with the dose of asbestos (principally amphiboles), but the exact shape of the relation is not known.

The experience of Canadian uranium miners is discussed to contrast the dose–response relations for fatal industrial accidents and industrial lung cancer in this population of workers.

Introduction

Recently on entering a hospital at which we work, one of us (E.D.A.) was confronted with a notice suspended across one of the corridors. It was inscribed with the single word 'asbestos', together with a skull and crossbones. Keep away or you may die, regardless of the dose, the notice seemed to say. An ancient hot-water supply system lagged with crocidolite was being removed. The notice was reminiscent of one that used to be seen at the entrance to certain beaches after World War II: 'Minefield. Danger of Death', also with the skull and crossbones. In this conference we are concerned with the assessments and perceptions that have led to the placing of such notices. In our paper on dose–response relations from epidemiological studies, we shall be taking most of our examples from studies of cancer, and because many of the factors so far identified that are known to increase the risk of cancer in man have been identified in industry, most of our examples will be cancers of occupational origin.

Quantitative measurements of personal exposure to a carcinogenic factor in human populations are relatively rare. Setting aside tobacco, alcoholic drinks and certain drugs, there are only two factors for which quantitative data about personal exposure and incidence of cancer are available over a range wide enough to make it possible to study the shape of the relation: chrysotile asbestos and ionizing radiation. Quantitative data about exposure to polycyclic hydrocarbons in gas workers are available that show that exposure to concentrations of 3,4-benzopyrene on average 100 times greater than that in London air did not quite double the risk of bronchial cancer compared with gas fitters working in the streets of the town (Lawther *et al.* 1964). However, sufficient data were not available to describe the shape of the dose–response relation.

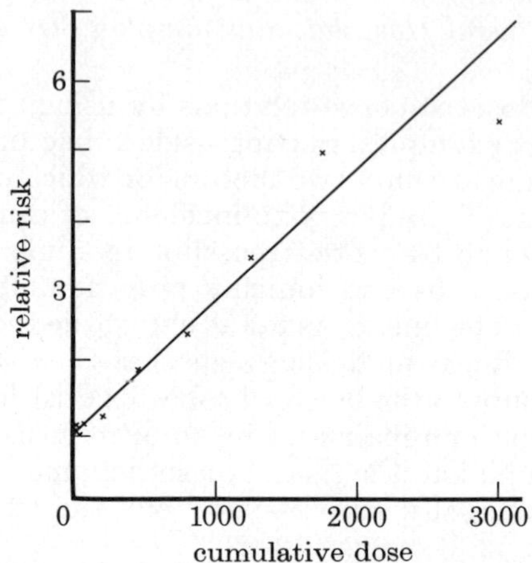

FIGURE 1. Dose–response relation for lung cancer in Quebec miners and millers: internal comparison between cases and controls (Liddell *et al.* 1977). Cumulative dose is measured in millions of particles per cubic foot year.

ASBESTOS

As far as asbestos is concerned, by far the largest quantity of data relating exposure to the risk of lung cancer in quantitative terms is contained in a series of studies of Quebec chrysotile miners and millers (Liddell *et al.* 1977). These studies have three important advantages: scale (more than 11000 men are involved); duration of follow-up (more than 4000 of the men are known to have died since 1935) and the wide range of data on cumulative dose of fibre that individual men have inhaled, including a substantial amount of data about low doses. The disadvantage of the material, which is common to all historical quantitative work in industry, is that the information about dose is an approximation summed from data recorded in different times and places with widely varying techniques and different units of measurement.

A plot of the dose–response relation for lung cancer among the Quebec workers is shown in figure 1. Cumulative asbestos dust exposures for 215 men who died of lung cancer, and for 1075 controls, are shown against relative risks calculated in relation to the lowest exposure group. A formal test for linearity is found to be highly significant ($\chi^2_{(1)} = 37.9$, $p < 0.01$) and the deviations about linearity insignificant ($\chi^2_{(8)} = 0.9$, $p > 0.1$). The authors and others have interpreted these data as suggesting that, for practical purposes, there is no 'safe threshold' in relation to chrysotile dust and lung cancer.

The acceptance of the linear relation has at least one important practical implication. If the excess risk of lung cancer is directly proportional to the cumulative dose of fibre at all levels of dose, it serves no good purpose to advise that work people should be moved to less dusty jobs after a period of exposure and be replaced by new workers. The yield of attributable cases of cancer will remain the same. On the other hand, if the index in the equation of the dose–response relation were greater than unity, as, for example, in a quadratic relation, turning-over the work force in the dustiest jobs would be expected to reduce the incidence of lung cancer.

TABLE 1. NUMBER OF EXTRA DEATHS FROM LUNG CANCER AMONG 10^6 BIRTHS FROM STATED EXPOSURE TO CHRYSOTILE ASBESTOS CONTINUOUSLY OVER 50 YEARS

(Adapted from table 26, Acheson & Gardner (1979).)

	place of exposure	
source of data for extrapolation	out of doors (0.01 fibre-years/ml)	in buildings (0.25 fibre-years/ml)
Quebec miners and millers	0.5	12.5
Rochdale factory workers	1.4	34

Knowledge of a dose–response relation, such as is indicated in figure 1, makes it possible to extrapolate to the situation where very large numbers of persons are exposed to extremely small doses of chrysotile fibre, e.g. those living or working in buildings where asbestos has been used in construction or out of doors in cities (Acheson & Gardner 1979). By using such data as are available concerning chrysotile levels in buildings containing asbestos and out of doors in urban environments in England, and by assuming a linear relation, it is possible to calculate the number of extra deaths from lung cancer among 10^6 people if they were exposed continuously to these levels for 50 years (Byrom *et al.* 1969; Rickards 1973). These are set out in table 1. This extrapolation shows that the number of extra cases of lung cancer that would be associated with pollution of buildings and the urban environment with chrysotile asbestos at the stated levels is very small indeed. Even so, the estimates overstate the real risk because the buildings chosen for survey had recently been completed or repaired and the dust levels within them were likely to decline. The figure for the concentration of chrysotile dust in the outside air used in the calculation was the highest recorded in England, and the great majority of the population would rarely, if ever, be exposed to such levels. By comparison with these figures more than 5×10^4 deaths among 10^6 persons will be due to lung cancer, and the majority of these will be associated with smoking. We are, of course, aware that there is a view that such extrapolations from industrial data recorded in the past should never be undertaken; in other words, that no data are better than indifferent data.

In studies of industrial workers exposed to asbestos in the past, lung cancer is only one of several causes of excess mortality, others being asbestosis and, particularly in relation to exposure to amphibole asbestos, mesothelioma of the pleura and peritoneum. Although about 300 cases of mesothelioma are recorded in England and Wales each year, a number that probably falls substantially short of the cases of lung cancer in which asbestos plays a part, mesothelioma seems to be the tumour that the public particularly fear in relation to asbestos exposure. This is probably for two reasons. First, although mesothelioma is rare, asbestos accounts for a high proportion of the cases, perhaps as many as two-thirds, and in this country asbestos is so far the only known causal factor. Secondly, it may be that because such cases are so easily perceived it is possible to identify the few in which apparently trivial exposures – very short periods of work or exposure at home to soiled clothes brought home from work or to dust in the neighbourhood of factories and dumps – have caused subsequent tumours. This has given rise to the view that for mesothelioma and amphibole asbestos the risk is constant at a wide range of dose levels. Although data about measured doses are not available, information from miners show that the mortality from mesothelioma increases with increasing intensity of exposure or duration of work (M. S. T. Hobbs, B. Murphy, A. W. Musk, J. Elder & F. Heyworth, unpublished, 1978); in other words, that, as with all other known carcinogens, smaller doses are less dangerous than larger doses.

Asbestos differs from other carcinogens in that its fibres are virtually indestructible at ordinary temperatures and pressures, and it may be that the cases of mesothelioma reported in relation to apparently small doses of dust represent the rare consequence of a very large number of trivial exposures.

Uranium miners

Uranium miners have also been the subject of studies that have enabled hypotheses to be made about the relation of dose (on this occasion radiation associated with radon daughter products) to mortality from lung cancer. In a study of 81 lung cancer deaths among more than 15000 uranium miners followed from 1955 to 1974, Hewitt concluded that a linear hypothesis provided an excellent fit and that there is no statistical justification for fitting any more complex model (Hewitt 1978). However, he also concluded that neither the data available, nor any other finite quantity of data, would exclude the possibility of a safe threshold. As the existence of a safe threshold at some level or other is a hypothesis that cannot be disproved, he prefers to introduce the concept of *highest plausible threshold* on the basis of the frequency distributions of exposure of subjects who have died of lung cancer and subjects who have survived. In the case in point the highest plausible threshold is a total dose of 60 working level months.

Although the primary object of the Ontario enquiry was to measure the deleterious effects of ionizing radiation in uranium miners, the epidemiological work has in fact identified a much more important industrial cause of excess mortality in these men. Table 2 shows the net excesses and deficiencies in the numbers of deaths in the miners for the main causes, compared with the experience of Ontario men of similar age.

At the time at which the mortality of the cohort was studied, accidents at work were found to contribute a much larger excess of deaths than irradiation. Furthermore, the accidents occur in younger men than the lung cancer cases. When this is taken into account, the deprivation of life-years by excessive fatal accidents was found to be eight times that resulting from the excess of lung cancer deaths. As the cohort ages, and more cases of lung cancer occur, the relative importance of deaths from accidents will decline.

TABLE 2. EXCESS DEATHS IN ONTARIO URANIUM MINERS BY MAJOR CAUSES (Hewitt (1978).)

violent causes	—	+188
motor vehicle accidents	+ 16	—
suicide	− 2	—
remainder (principally accidents at work)	+174	—
natural causes	—	−137
cancer of lung	+ 36	—
cardiovascular	− 93	—
other	− 80	—
all causes	—	+ 51

Apart from the obvious point of the importance of seeing risks in perspective in relation to all other ascertainable risks, two further points may be made about this finding. The first is that any measure directed at reducing the length of service (and therefore of irradiation) in the mine, which increases the number of young and inexperienced miners, may have the opposite effect to that intended and increase, rather than reduce, the overall excess mortality attributable to the work, at least in terms of deprivation of life years. The second is to ask, does the study of accidents in the mines and their prevention receive as much attention as the prevention of irradiation-induced cancer? And if not, why not?

PERCEPTION

With this question we shall turn from the assessment of risk to its perception, drawing on the examples already quoted, well aware that we are moving from data, however crude, to opinion or even conjecture. In the field of cancer, the two elements of perception of risk that are essential if society is to move to a rational policy for prevention that balances costs and benefits, and which are lacking today, are a view of a particular risk in the perspective of all other risks and the knowledge that all cancer risks so far disclosed are dose-related. But of course there are many other factors involved in how a risk is perceived, including the extent to which the size of the risk is seen to have been disclosed, the element of choice involved (the risk associated with a hobby compared with work), and fear and ignorance about the nature of the consequence. A number of other points arise in the contrast already drawn between deaths from accidents in the uranium mine and deaths from radiation-induced lung cancer. In the former the consequence of the accident is immediate, and the risk ceases at the factory gates and is extinguished by leaving employment. In the latter, for both asbestos and uranium the effect is delayed and a risk remains long after exposure has ceased and even after retirement from active work. There remains the question of the perception of risks associated with habits, such as smoking, associated with gratification. This is relevant because in relation to exposure to both asbestos and uranium the effect of smoking is synergistic, and possibly multiplicative. This introduces a further element of emotion, which colours perceptions and tests attitudes. Suggestions that smoking should be forbidden in asbestos factories have not been widely accepted, and yet such prohibition has long been accepted where there is a risk of fire or explosion. Perhaps arrangements should be made to ensure that in asbestos factories it is no more unsafe to smoke on the shop-floor than in the boardroom.

In conclusion, there are at least four points that should be taken into account when a decision is taken to reduce the dose of a carcinogen to which a population is exposed. The first is that the incidence of the related cancer will be reduced among the persons whose dose is reduced, although the benefit to the population currently exposed will be less than to future exposed populations. Secondly, there is at least a possibility that the means chosen to reduce the risk, either by changing work patterns or introducing a substitute, may introduce other risks. A third point to be considered is whether by reducing a risk in one population one increases it in a different one. Thus, if regulations were made that ruled out the use of chrysotile in brake linings, this might reduce the risk of lung cancer in production workers but would almost certainly increase the mortality in the general population from road accidents. The final consideration that must always be taken into account is the economic cost of reducing risks.

References (Acheson & Gardner)

Acheson, E. D. & Gardner, M. J. 1979 *Asbestos*, vol. 2 (*Final report of the advisory committee*). London: H.M.S.O.

Byrom, J. C., Hodgson, A. A. & Holmes, S. 1969 *Ann. occup. Hyg.* **12**, 141.

Hewitt, D. 1978 In *Report of the Royal Commission on the health and safety of workers in mines*, pp. 78–83 and Appendix C. Government of Ontario.

Lawther, P. J., Commins, B. T. & Waller, R. E. 1965 *Br. J. ind. Med.* **22**, 13–20.

Liddell, F. D. K., McDonald, J. C. & Thomas, D. C. 1977 *Jl R. stat. Soc.* A **140**, 469.

Rickards, A. L. 1973 *Justus Liebigs Annln Chem.* **45**, 809.

Discussion

H. S. Dobbs (*Royal National Orthopaedic Hospital, Stanmore, U.K.*). Professor Acheson mentioned two studies in which the dose–response relation was linear. Could he also give examples of studies in which a nonlinear, possibly quadratic, relation exists?

E. D. Acheson. I know of no examples where a nonlinear dose–response relation has been generally accepted in relation to the action of industrial carcinogens in man. It is possible that there are some examples of nonlinear dose–response relations in respect of the pharmacological effects of drugs in man. However, this is outside the terms of reference of our paper.

Proc. R. Soc. Lond. A **376**, 87–101 (1981)
Printed in Great Britain

Quantification of risk in medical procedures

By E. E. Pochin
National Radiological Protection Board, Harwell, Didcot, Oxfordshire OX11 0RQ, U.K.

In most medical procedures, the benefit appears likely to exceed the risk by so much that detailed quantification of the risk is unnecessary. In some instances, however, it is important to attempt to estimate and compare the likely amount of the risk and of the benefit, to determine whether, or when, the use of the procedure is justified. This need arises in the use of radiological screening programmes for the early diagnosis of certain forms of cancer, to assess the age above which such a programme would save more lives by making early diagnoses than it would lose by inducing cancer in the tissues irradiated. It is also informative to estimate the levels of risk that might be involved in research studies involving radiation exposure.

Conventional diagnostic and therapeutic procedures entail risks of fatality ranging over at least four orders of magnitude, and clearly relate to the urgency of the situations in which the procedures need to be used. Estimates of such risks are reviewed.

Introduction

The quantification of risk may be valuable in a number of ways, and here the word 'risk' is used to denote the estimated probability that a specified harmful effect will result from specified conditions or procedures. It is informative to compare, for example, the probability of certain types of harm resulting per year from work in different industries or sections of an industry, or per mile travelled by different means of transport, or per hour engaged in different sporting activities, or per unit electrical output derived from different primary fuel cycles (Pochin 1974, 1980*a*).

In none of these cases does the estimated risk alone determine the acceptability of an activity or the choice between alternatives, but, in all of them, the size of the objective risk must presumably be regarded as a significant component in such decisions, to be reviewed alongside the character of the types of harm involved, and many other factors affecting the decision. Whatever assessments may be made of the perceived risk of the alternatives, however, it is a public health responsibility to know and consider the objective risk.

The risks of medical procedures are as difficult to quantify in any exact and rigorous way as are the risks in other fields. They have the same usefulness, however, in indicating the often widely different orders of magnitude of the hazards of different procedures, and in posing in numerical form the question of whether the benefits from a procedure clearly outweigh the risks, or whether the benefits need to be reviewed in a similarly careful and quantitative fashion.

Radiological screening programmes

An immediately important problem, in which a numerical risk–benefit analysis is not only necessary but also practicable, arises in the use of radiological screening of sections of the population to detect forms of cancer at early stages, when they can be more successfully treated than when symptoms have already developed. This question has been closely studied in connection with breast cancer, for which there is increasing evidence of the frequency with which previously unidentified cancers can be detected by such screening programmes at different ages, and also of the likely frequency with which breast cancers might be induced (Land *et al.* 1980) by the irradiation of the breast involved in the examination. More needs to be known of the way in which this frequency of detection varies with the radiological techniques adopted and hence with the radiation dose delivered, and also of the types and curability of the cancers so detected in comparison with the curability of breast cancers developing in unscreened populations. It is evident, however, that the benefit of such screening increases progressively with age, owing to the greater incidence of breast cancer in older women, and that the risks of screening will decrease with age, owing largely to the long average interval, of 20 years or more, between radiation exposure and the appearance of induced cancers. The available information has suggested that, at least above the age of 50, more deaths from breast cancer are likely to be prevented by efficient techniques of screening than are induced by them (Breslow *et al.* 1977).

A similar investigation has been made into the risk–benefit balance of radiological screening for stomach cancer in Japan, where this form of malignancy is responsible for over half of the total cancer mortality. This is an important type of cancer to detect by mass survey of apparently healthy people, since it is one that is very liable to prove fatal unless detected before symptoms develop; X-ray examination offers the only practicable method of screening on a mass scale.

The quantification of this study was greatly aided by the fact that such screening has been practised in Japan since 1960, and as many as four million persons were examined in this way in 1978. Reliable data on benefit were therefore available, with observations on the frequency with which otherwise undetected stomach cancers were found by the techniques of screening used. Iinuma *et al.* (1979) had good evidence also on the frequency with which cancers so detected were cured, and how frequently they caused death despite being so detected; data were obtained on the variation of all these figures with age and sex. This information was compared with the incidence and mortality of stomach cancer in the absence of a screening programme. The net benefit from screening was expressed in terms of the average number of years of life expectancy lost owing to deaths from stomach cancer in a screened population, compared with the greater number of years lost, owing to later diagnosis, in the unscreened population (table 1).

The corresponding estimates of risk were derived on the basis of the increased average radiation exposure of people having regular radiological screening, compared with that from any stomach X-rays received in the absence of such a programme. Estimates were made of the average absorbed doses delivered to

TABLE 1. RADIOLOGICAL SCREENING FOR STOMACH CANCER (IINUMA *ET AL.* 1979)

(Years of life lost, with or without screening (per 1000 persons).)

age (years) ...	20–24	25–29	30–34	35–39	40–44	45–49
with screening						
from stomach cancer	0.24	0.44	0.87	1.38	2.19	4.14
from induced cancers	1.48	1.20	0.92	0.67	0.47	0.31
total years lost	**1.72**	**1.64**	**1.79**	**2.05**	**2.66**	**4.45**
without screening						
from stomach cancer	0.60	1.12	2.23	3.52	5.57	10.53
from induced cancers	0.00	0.01	0.01	0.01	0.01	0.02
total years lost	**0.60**	**1.13**	**2.24**	**3.53**	**5.58**	**10.55**
net benefit from screening, by difference						
males (years per 1000)	−1.12	−0.51	+0.45	+1.48	+2.92	+6.10
net benefit from screening (derived similarly)						
females (years per 1000)	−1.85	−0.98	+0.52	+2.03	+3.11	+4.49

bone marrow and to the body organs mainly exposed during the examinations. They were combined with the values available for the frequency with which leukaemia or cancer is found to be induced, per unit radiation exposure of these organs. Considerable uncertainty is necessarily involved in making the risk estimates for some of the organs exposed, and also in the distribution of the time intervals assumed between radiation exposure and the development of any induced malignancies, and therefore the consequent loss of life expectancy. The calculated net detriment or benefit from screening, however (table 1), appears to vary rather rapidly with the age at which screening is carried out. The estimate of the age above which a net benefit should be obtained is therefore not likely to be highly sensitive to the losses of life expectancy attributed to radiation-induced malignancies. Indeed the age – about 30 years – at which risk and benefit are judged to be in balance on these criteria, would only be raised or lowered by 4 years if the estimates of radiation risk were doubled or halved (in males, and by slightly less in females).

Although no claim can be made, or is made, by the authors of this study, that the evaluations either of risk or of benefit are exact, even the approximate conclusions reached must be regarded as important. In the absence of any such estimates of the value of this form of screening, and of the age above which it should be used, no valid judgement could be made on purely intuitive grounds. Moreover, to obtain direct epidemiological evidence on these questions would require a series of screening programmes, starting at different ages, continuing for several decades until all radiation-induced cancers had developed, and involving the very large populations needed to provide any statistically reliable estimate of the excess number of such induced cancers above the natural cancer expectation.

Investigations with radiological techniques

Even when no numerical estimate is available on the likely benefit expected from a procedure, it may still be valuable to estimate the possible levels of risk entailed. Such an estimate may make it evident that the amount of benefit will greatly exceed, or be exceeded by, the amount of risk; or it may stimulate investigation to obtain at least some quantification of the benefit.

Considerations of this sort are involved in the use of radiological techniques for investigating human disease or metabolism. In most of the conventional diagnostic uses of radionuclides or radiopharmaceuticals in nuclear medicine, it is known, for any administered activity, what absorbed dose of radiation is normally delivered to body tissues as a whole, or to those organs in which the radionuclide is selectively concentrated and retained. Moreover, the doses delivered by 'cross-fire' from these to other organs have, in many cases, been assessed on a systematic basis. The variations in these levels of absorbed dose have also been examined in various forms of disease in which the uptake or turnover of the nuclide may be altered as a result of tissue damage or metabolic abnormality.

An estimate can thus be given, for any radionuclide test, either of the absorbed doses delivered to body tissues as a whole if they are irradiated uniformly, or to the various body organs if they are irradiated selectively. In the latter case it is convenient to weight the different organ doses according to the estimated risk of induction of major harmful effect, whether of fatal cancer or of substantial genetic abnormality, by irradiation of each organ. In this way the total of all organ doses can be expressed as the equivalent uniform whole-body dose that would carry the same risk of harm as that from the selective irradiation of the various organs. This estimate of an 'effective' whole-body dose is obtained by the use of organ weighting factors (I.C.R.P. 1977*a*) based on the risk observed in epidemiological studies of the induction of malignancies or of mutations following exposure of individual organs or tissues. A similar method has been used to compare the possible effect of different types of diagnostic X-ray examination, which irradiate parts of the body of differing sensitivity, by means of the corresponding 'effective dose' of whole-body radiation.

The effective doses resulting from diagnostic examinations vary, and estimates are given in table 2 for conventional types of X-ray and radionuclide tests. For the latter, the effective doses have an approximately log-normal distribution, with a geometric mean of about 0.4 mSv for tests of different types carried out with the

TABLE 2. EFFECTIVE DOSE FROM DIAGNOSTIC PROCEDURES

equivalent whole-body 'effective dose' (mSv)	number of diagnostic tests by X-rays	radionuclides
0.0001–0.001	—	1
0.001–0.01	—	3
0.01–0.1	1	21
0.1–1.0	18	33
1.0–10	14	25
> 10	—	5
geometric mean dose	0.7 mSv	0.4 mSv

maximum usual administered activity of radionuclide. This value is comparable with that estimated for a range of diagnostic X-ray examinations, with a mean effective dose of 0.7 mSv. For comparison, it may be noted that natural sources of radiation deliver an annual effective dose of about 1 mSv (UNSCEAR 1977).

The uses of radiation and radionuclides in medical research were considered recently by an Expert Committee of the World Health Organization (W.H.O. 1977). Three categories of research project were distinguished, according to whether the effective dose (or future dose commitment) delivered in any year was 0.5 mSv or less, or lay between 0.5 and 5 mSv, or between 5 and 50 mSv. Greater exposures were considered likely to be acceptable only in special circumstances.

The exposures involved in the lowest category would be within the normal range of variation in radiation received annually from natural sources, or from houses of different building materials. The exposures in W.H.O.'s category II are within the annual limit recommended for any member of the public from all man-made (but non-medical) sources. Those in category III are within the limit recommended for occupational exposures.

What, then, are the risks from an investigation involving a dose of, say, 5 mSv (i.e. at the boundary between W.H.O.'s categories II and III, and exceeding that attributable to most conventional X-ray or radionuclide examinations)? It can be estimated that the average risk of a substantial genetic abnormality in the descendants of the irradiated individual would fall from about 10^{-4} for such an exposure below the age of 18 years, to rather less than half this value for exposure at age 30 years, and to below 10^{-5} after the age of 40 (in males; or 35 in females). The risk of inducing a fatal subsequent malignancy in the individual exposed would fall slowly with age, from about 6×10^{-5} for exposure below the age of 25, to half this value for exposure at age 55 (I.C.R.P. 1977*b*). For a man aged 55, this risk is equal to that of his dying from natural causes on any one day; or, for the man aged 25, in any one month.

A typical large investigation of metabolism or organ function might involve studies on 25 people. The total risk from such a study, requiring effective doses of 5 mSv to each, would vary with the age of the subjects. If, for example, all were of age 30, it would be estimated that either one death, or one substantial genetic abnormality, would result from every 400 such investigations. (An equal risk would result from a total of 700 such studies at age 40, or 1000 at age 50.)

Any risk estimates of this sort must necessarily be highly tentative and approximate. It must surely be regarded as helpful to any prospective volunteer, however, as well as to the community and to the investigator, to offer some 'feel' for the order of magnitude of his possible risk, which in this instance would correspond to a probability of subsequent death from cancer being increased from 0.22 to 0.2201. And it could be helpful also to know that the investigation would be worth while, in a crude overall arithmetic estimate of lives saved and lost, if its results had a 1 % chance of saving even one life, world wide and in the course of all time.

Orders of magnitude in medical risks

It is to be expected that, as the urgencies of diagnosis and of treatment differ widely in different medical conditions, so the risk that could properly be regarded as acceptable, if necessary, in diagnostic or therapeutic procedures might vary accordingly. In this sense, as in all other risk–benefit analyses, the amount of benefit should set an upper limit to the justifiable amount of risk.

The benefits derived from medical procedures are usually as poorly quantified, and often as difficult to quantify, as are benefits in other fields. The risks, however, can commonly be estimated, at least in terms of some limited criterion such as the probability of death attributable to the procedure, and when so quantified prove to range over some five orders of magnitude for different conventional methods of diagnosis or treatment.

Safety of medicaments

Thus Girdwood (1974) reviewed reports to the Committee on the Safety of Drugs (or, later, of Medicines) of the number of deaths during a 10 year period that were regarded as being possibly due to different medicaments. He compared these figures with the number of prescriptions issued annually by general practitioners, and so estimated the number of deaths per 10^6 prescriptions that might be due to the use of different drugs (excluding known instances of overdosage). He emphasized, however, that not only the number of deaths, but also the number of courses of treatment, were likely to have been underestimated, and that the 'mortality rates' observed were more reliable as indicators of the relative, rather than of the absolute, risk of different drugs. Moreover, in some cases no valid rate could be derived, either because the drugs could be obtained without prescriptions or because they were used predominantly in hospitals, without prescription from general practitioners.

Of over 200 types of drug or preparation for which prescription rates were obtainable, 20 had mortality rates of 1 or more deaths per 10^6 prescriptions. In one of these the estimated rate was about 150, while the rates for the remainder ranged from 1 to 18 deaths per 10^6 prescriptions.

Vere (1976) made a similar analysis of the frequency of deaths from haematological effects (agranulocytosis and aplastic or haemolytic anaemias). The mortality rates for the 16 drugs evaluated in this way had a median value of 0.5 per 10^6 prescriptions, the rates having values of less than 1 in 11 cases, between 1 and 10 in 4, and a value of 20 in one case.

These data, despite their approximate nature, suggest a risk of death per course of treatment in the order of 10^{-6} or less for most medicaments in common use, although with occasional instances at 10^{-5} or 10^{-4}. In the early stages of the use of a new drug, however, the risk may evidently be substantially higher. Such risks will often be unquantified, if the drug is withdrawn from use on the first reports of danger in its clinical trials. With thiouracil, however, a thiourea derivative that was effective in decreasing thyroid overactivity in thyrotoxicosis, three reports (Moore 1946; van Winkle *et al.* 1946; Fowler 1946) had appeared within 3 years of the drug's introduction to clinical use, and had recorded a total of 33 deaths, from agranulocytosis, in 8409 patients treated with the drug: an average mortality risk of 4×10^{-3} per treatment.

A similarly high risk is suggested also by the death rates of patients treated with clofibrate to lower the level of the blood cholesterol (Oliver *et al.* 1980). The age-standardized annual death rate within the first 10 years from the start of a 5 year course of treatment was 8.1 ± 0.4 s.e. per 1000 in patients so treated, compared with 6.6 ± 0.4 and 4.8 ± 0.3 in patients with, and without, raised cholesterol values, but not treated with clofibrate. These data also therefore indicate an excess risk of rather over 10^{-3} per treatment.

There is clearly scope for a similar careful numerical evaluation of the safety of other drugs, and a corresponding assessment where possible of the value of a drug in the treatment of serious disease, when a substantial risk is detected and when no safer treatment is available.

Safety of vaccinations

The frequency with which death may result from the prophylactic vaccination of healthy people is likely to vary with a number of factors, such as the type of vaccination and the age at which it is performed. Overall, however, a low fatality risk in the order of 10^{-6} is suggested by the average rate of 3.3 deaths per year between 1967 and 1976 in England and Wales (O.P.C.S. 1977, 1979*a*) that were attributed to the effects of vaccinations (under I.C.D. categories E933 and 934), during a period in which an average of 3.7×10^6 vaccinations were carried out each year (*Social trends* 1979) (against the eight types of infective disease for which vaccinations were predominantly made).

Such estimations of risk, both of death and of any other serious effect, for different types of vaccination have obvious importance in the risk–benefit assessment of whether, at any given time, the risks of vaccination, low as they are, may exceed the benefits obtained in the prevention of disease (Cutting 1980). This decision, however, requires as good an estimate of the benefit as of the risk, and assessments are needed both of the frequency and mortality or other effects of the disease without vaccination, and of its frequency and effects despite vaccination. The occasional fatal effects of smallpox vaccination (du Mont & Beach 1979), however, illustrate the risk–benefit imbalance in any continued vaccination, following the elimination of this disease. The action of the World Health Organization must incidentally be regarded as one of the most cost-effective major operations ever staged. The total cost of smallpox eradication is stated by W.H.O. (1980) to have been about U.S.\$ 300×10^6 and to be saving an estimated 2×10^6 lives per year. Whatever cash value might be assigned to a human life, it must certainly exceed the sum of \$15 implied by a 10% annual interest on the capital sum expended.

Radiological diagnosis

The diagnostic X-ray and radionuclide procedures that were reviewed above involved a range of 'effective' dose equivalents with (geometric) mean values of some tenths of a millisievert (table 2). The risks corresponding to these exposures would vary, both for effects in the individuals and in their descendants, with age and with sex. As an average throughout the population, however, the risk of induction of a fatal cancer has been estimated as in the region of 10^{-5} per millisievert, with a similar risk of curable cancer (of certain body organs, particularly of the thyroid) and with a somewhat smaller risk of major genetic abnormality. The risk of death or of major genetic effect from the mean doses, of 0.7 and 0.4 mSv in the two types of radiological procedure, would therefore be in the region of 10^{-5} per examination. The estimated absorbed doses from different types of radionuclide test are distributed round their mean value with a (logarithmic) standard deviation of about 10, so that the levels of risk may vary similarly. For X-ray examination, the variation appears to be less, but estimates based on average values for a given investigation do not take account of the substantial variation (Adrian 1960) still observed (Wall *et al.* 1980) in the dose delivered during the performance of the same examination in different conditions; this source of variability will certainly be less in any radionuclide test. The overall variation of both types, of radiological and nuclear medicine diagnosis, therefore imply a mean risk of 10^{-5}, varying commonly with different tests between 10^{-6} and 10^{-4}.

These estimates assume a linear proportionality between size of dose and frequency of effect which, for these radiations of low energy transfer, may overestimate risk (B.E.I.R. Committee 1980). An increased incidence of breast cancers was, however, observed in patients who had required repeated fluoroscopic examinations during the lung collapse therapy of their tuberculosis, and who had been facing towards the X-ray tube during the examination (Boice & Monson 1977). Here the risk of fatal breast cancer is likely to have been in the order of 10^{-4} per examination, and was detectable epidemiologically on account of the large number of examinations needed. A similar risk of fatal malignancy, probably of about 2×10^{-4}, was indicated by the observations of Stewart & Kneale (1970) on the occurrence of leukaemia and other cancers in children who had been exposed *in utero* during pelvic X-ray of their mothers.

Moreover, as with new drugs, the introduction of new radiological techniques may sometimes be recognized as involving high risk. Thorotrast, a suspension of thorium oxide, was formerly used as 'contrast medium' for outlining the distribution of blood vessels after its injection into them, the high atomic number of the thorium giving good contrast on X-ray. The subsequent retention of the radioactive thorium, particularly in liver and bone, was, however, recognized to be associated with a considerable induction of liver cancers and of leukaemia, with a mortality rate estimated as 5×10^{-2} from the former malignancy and 1.2×10^{-2} from the latter, from a typical (25 ml) injection (UNSCEAR 1977).

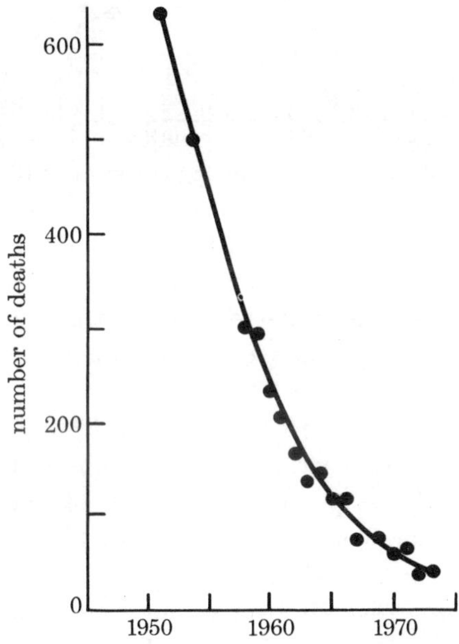

FIGURE 1. Number of deaths attributed to anaesthetic per 10^6 operations (Office of Health Economics 1976).

Liver biopsy

The technique of obtaining samples of liver tissue in the diagnosis of hepatic disease was made safer in the 1960s by the use of the Menghini needle, and several studies indicate the extent to which the risk of this procedure was reduced. The survey by Zamchek & Klausenstock (1953) before the introduction of the Menghini technique recorded 34 deaths after 20000 needle biopsies of the liver. Those by Thaler (1964) and by Lindner (1967) after its introduction, however, together record 16 deaths after about 100000 biopsies, a fall in risk from 2×10^{-3} to 2×10^{-4}. These, however, are estimates averaged over a very heterogeneous

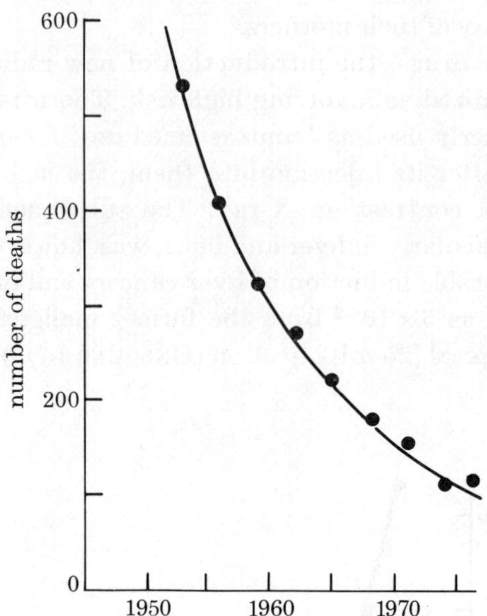

FIGURE 2. Number of maternal deaths, per 10^6 maternities (O.P.C.S. 1979*b*).

population at risk, since the hazards of such biopsies depend considerably on the likelihood of bleeding following the puncture, and hence on the degree of jaundice, blood coagulability, and portal vein hypertension; just as the benefits or urgency for accurate diagnosis will also vary from patient to patient.

Anaesthetic and maternal deaths

The frequency with which death is attributable to anaesthetic used in major surgical operations will clearly vary with the length of operation, the state of health of the patient and other factors. The Office of Health Economics (1976), however, quotes values for the overall average rates of death attributed to the anaesthetic, and these fall progressively from about 6×10^{-4} in 1951, to 4×10^{-5} in 1973 (figure 1), indicating a halving of the rate about every 6 years during this period.

The risk of maternal death at childbirth has fallen rather more slowly (figure 2) from a similar figure, of about 5×10^{-4} per maternity in 1953, to 10^{-4} in the mid-1970s (O.P.C.S. 1979*b*), while the maternal risk per legal abortion in England and Wales (Lewis 1980) is estimated to have fallen from rather over 10^{-3} in 1967 to 4×10^{-5} in 1973 and subsequently (figure 3).

Risks in therapy

The risks involved in the treatment of serious disease, whether by medical, surgical or radiological techniques, are clearly – and properly – likely to have varied with the severity or risks of the condition requiring treatment. They have,

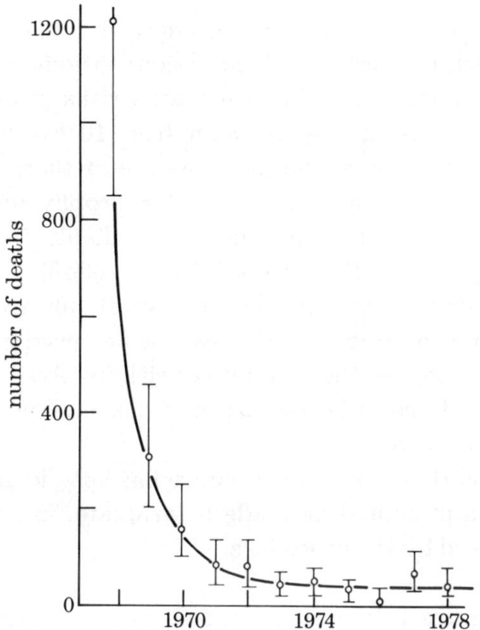

FIGURE 3. Number of maternal deaths, per 10^6 legal abortions with 90 % confidence limits; (Lewis 1980).

however, been particularly closely studied in regard to the radiotherapy of benign conditions, since any excess of cancer incidence in the organs or regions irradiated allows an estimate of the carcinogenic risk of radiation exposures, and since the long survival of patients treated for such conditions makes the observation of such late effects possible.

In this way the treatment by radiation of the disabling condition ankylosing spondylitis was found by Court Brown & Doll (1965) to be followed by an increased mortality of about 1 % due to malignancies arising from the heavily irradiated areas, and later studies (Smith & Doll 1978) of the cancer and leukaemia mortality of patients receiving single courses of this treatment confirmed this figure.

The treatment of the same condition by internal irradiation, from radium-224 administered intravenously as Peteosthor, was followed by an equal excess of malignancies (Spiess & Mays 1970), the tumours consisting in this case of bone sarcomas, most proving fatal. The overall incidence of 49 sarcomas in 925 patients, however, who had received such injections for all conditions, before the treatment was largely discontinued in 1951, corresponds to an incidence rate of 5.3 % and a fatality rate that is likely to have exceeded 4 %.

Conclusions

As in other fields of interest in which the numerical values of risks have been determined and compared, the risks involved in conventional medical procedures vary over a remarkably wide range. The estimated risks of death from past or present diagnostic methods thus appear to range from 10^{-6} or less for many types of drug treatment, to 6×10^{-2} from the previous use of thorotrast as a contrast medium for intra-arterial injection. Similarly, for prophylactic or therapeutic measures, a range of five orders of magnitude is identifiable, from the current use of vaccinations to the past use of Peteosthor (Pochin 1980*b*).

Such numerical estimates of risk may be of interest but are of little value in isolation. However, when the same objective can be achieved by different means, it becomes important to compare the amounts of risk involved in the use of these different means, as well as to consider the types of risk entailed and the weighting that should be attached to them.

It must be of value also that, when a procedure has been identified as involving substantial risk, an attempt should be made to evaluate, in any commensurable terms, the benefits achieved by the procedure.

Tables 1 and 2 and figures 1–3 are reproduced, with modifications, from *Risk–benefit analysis in drug research* (Lancaster: M.T.P. Press), by permission of the editor, J. F. Cavalla.

References (Pochin)

Adrian 1960 *Radiological hazards to patients: second report of the Committee*. Ministry of Health, Ministry of Health for Scotland. London: H.M.S.O.

B.E.I.R. Committee 1980 *The effects on populations of exposure to low levels of ionizing radiation*. Report of the Committee on the Biological Effects of Ionizing Radiation. Washington D.C.: National Academy of Sciences.

Boice, J. D., Jr. & Monson, R. R. 1977 *J. natn. Cancer Inst.* **59**, 823–832.

Breslow, L., Thomas, L. B. & Upton, A. C. 1977 *J. natn. Cancer Inst.* **59**, 467–541.

Court Brown, W. M. & Doll, R. 1965 *Br. med. J.* ii, 1327–1332.

Cutting, W. A. M. 1980 *Lancet* ii, 634–635.

du Mont, G. C. L. & Beach, R. C. 1979 *Br. med. J.* i, 1398–1399.

Fowler, E. F. 1946 *Int. Abstr. Surg.* **83**, 313–322.

Girdwood, R. H. 1974 *Br. med. J.* i, 501–504.

Iinuma, T. A., Tateno, Y., Umegaki, Y. & Hashizume, T. 1979 In *Radiation Research: Proceedings of the Sixth International Congress of Radiation Research* (ed. S. Okada *et al.*), pp. 980–988. Tokyo:

International Commission on Radiological Protection 1977*a* *Ann. I.C.R.P.* **1** (3), 1–53. (*I.C.R.P. Publ.* no. 26.) Oxford: Pergamon Press.

International Commission on Radiological Protection 1977*b* *Ann. I.C.R.P.* **1** (4), 1–24. (*I.C.R.P. Publ.* no. 27.) Oxford: Pergamon Press.

Land, C. E., Boice, J. D., Jr., Shore, R. E., Norman. J. E. & Tokunaga, M. 1980 *J. natn. Cancer Inst.* **65**, 353–376.

Lewis, T. L. T. 1980 *Br. med. J.* i, 295.

Lindner, H. 1967 *Dt. med. Wschr.* **92**, 1751–1757.

Moore, F. D. 1946 *J. Am. med. Ass.* **130**, 315–319.

Office of Health Economics 1976 *Report No.* 55 *on Anaesthesia*. London: H.M.S.O.
Office of Population Censuses and Surveys 1977 *Mortality statistics of childhood and maternity*. DH3, no. 4. London: H.M.S.O.
Office of Population Censuses and Surveys 1979*a* *Birth statistics*. FM1, no. 3. London: H.M.S.O.
Office of Population Censuses and Surveys 1979*b* *Mortality statistics* 1977, *England and Wales*. Series DH 1, no. 5. London: H.M.S.O.
Oliver, M. F., Heady, J. A., Morris, J. N. & Cooper, J. 1980 *Lancet* ii, 379–385.
Pochin, E. E. 1974 *Community Health* **6**, 2–13.
Pochin, E. E. 1980*a* *Physics Med. Biol.* **25**, 1–12.
Pochin, E. E. 1980*b* In *Risk–benefit analysis in drug research* (ed. J. F. Cavalla). Lancaster: M.T.P. Press. (In the press.)
Smith, P. G. & Doll, R. 1978 In *Late biological effects of ionizing radiation*, vol. 1, pp. 205–218. (I.A.E.A.-SM-224/711.) Vienna: I.A.E.A.
Social trends 1979 Vol. 9 (ed. E. J. Thomson). Government Statistical Service. London: H.M.S.O.
Spiess, H. & Mays, C. W. 1970 *Hlth Phys.* **19**, 713–729.
Stewart, A. M. & Kneale, G. W. 1970 *Lancet* i, 1185–1188.
Thaler, H. 1964 *Wien klin. Wschr.* **29**, 533–538.
United Nations Scientific Committee on the Effects of Atomic Radiation (UNSCEAR) 1977 *Report to the General Assembly*. New York: United Nations.
van Winkle, W., Jr., Hardy, S. M., Hazel, G. R., Hines, D. C., Newcomer, H. S., Sharp, E. A. & Sisk, W. N. 1946 *J. Am. med. Ass.* **130**, 343–347.
Vere, D. W. 1976 *Proc. R. Soc. Med.* **69**, 105–107.
Wall, B. F., Fisher, E. S., Shrimpton, P. C. & Rae, S. 1980 *Current levels of gonadal irradiation from a selection of routine diagnostic X-ray examinations in Great Britain*. National Radiological Protection Board report no. NRPB–R105.
World Health Organization 1977 *Use of ionizing radiation and radionuclides on human beings for medical research, training and non-medical purposes: report of a W.H.O. expert committee*. Technical report series no. 611. Geneva: W.H.O.
World Health Organization 1980 *Wld Hlth*, May, p. 18.
Zamchek, N. & Klausenstock, O. 1953 *New Engl. J. Med.* **249**, 1062–1069.

Discussion

R. SCOTT RUSSELL (*East Hanney, Wantage, Oxfordshire, U.K.*). The estimates of risk from radiation, which Sir Edward Pochin discussed, may help us to understand some reasons why the public have difficulties in perceiving the risks that they might experience from the industrial use of nuclear power, and also suggest how a more balanced appraisal might be assisted.

Because the disposal of radioactive waste is much in the public mind I confine myself to this aspect. Official attitudes justify the assumption that no method of disposal would be judged acceptable if it portended annual radiation doses to members of the population that approached those from natural background, or were indeed more than a fraction of variations in that background between different parts of the country. On Sir Edward's basis the risk might therefore be estimated as about 10^{-6}. Would this statement help public comprehension? Doubts are at least encouraged by the indication, earlier in this discussion, that even those familiar with judging risks can find it difficult to perceive the implication of risks of this low magnitude. Beyond this we should take account of two other statements by Sir Edward. First, if guidance is based on estimates of risk, it is not adequate merely to say that it is very small; its magnitude should be stated. Secondly, all quantitative estimates of risk from very low radiation doses are subject to considerable uncertainty. This is an inescapable consequence of the fact that direct evidence on dose–response relations can be obtained only for much higher doses, which cause effects sufficiently frequently for accurate measurement. The prediction of effects in the background range thus involves extrapolation over several orders of magnitude; the resulting uncertainty will be evident to biologists.

Accordingly, despite the undoubted value of risk estimates when much higher doses are involved, there seem to be grounds for doubting whether, either in terms of public cognition or scientific reliability, these estimates are by themselves a sound basis for judging risks from very low radiation doses. It therefore seems surprising that much more reliance has not been placed on studies of the effects of background radiation, despite their limitations. The most careful scrutiny of the available data has not shown that variations in background radiation cause detectable harm. This is no proof of its entire absence; it may occur with a frequency too low for detection. None the less, it seems logical to conclude that doses less than a small fraction of background doses do not have consequences that could ever be perceived by members of the population. They are therefore 'safe' in the popular meaning of that word, but 'insignificant dose level' would be a more correct description. Would discussions of this type assist the public to a more valid perception of the effects on them, and on their descendants, of the disposal of radioactive waste with proper safeguards? As scientists, we would prefer more quantitative guidance, but is it justified by the facts that we now have or that we can expect from future research? Much evidence that must be assessed, to give adequate answers to these questions, lies outside the sphere of experimental science. Thus it seems very timely that, at the beginning of this meeting, Lord Ashby emphasized the contribution that the legal profession could give to discussions of risk.

E. E. Pochin. It would certainly be valuable if a study of any harm associated with different levels of background radiation gave useful direct information on the risk of low radiation exposure. In most cases, however, the size of populations examined, the difficulties of disease ascertainment in these groups, and the absence of strictly comparable comparison populations at lower radiation exposure, make it possible only to exclude risk rates that are substantially higher than seem likely on other grounds. A possible exception is the Chinese study reported in *Science* (**209**, 877–880 (1980)), which at least appears to exclude any cancer induction considerably higher than ordinarily estimated.

H. J. Dunster (*Health and Safety Executive, London, U.K.*). As a footnote to Sir Edward's reply, may I say that I find it difficult to distinguish, except in arithmetic terms, between probabilities of 10^{-4} and 10^{-5}. The existence of familiar voluntary risks provides absolutely no justification for improving other risks but can be used to explain otherwise unintelligible figures. An imposed, or offered, risk of say 4×10^{-6} can perhaps be better understood if it is translated as the risk of smoking four cigarettes.

Proc. R. Soc. Lond. A **376**, 103–119 (1981)
Printed in Great Britain

QUANTIFICATION OF PHYSICAL AND ENGINEERING RISKS

Experience in the quantification of engineering risks

BY F. R. FARMER†
United Kingdom Atomic Energy Authority, Wigshaw Lane, Culcheth, Warrington, WA3 4NE, U.K.

This paper concentrates mainly on the quantification of selected risks in industrial activities and with those events that might harm the public – possibly in large numbers – rather than those at work.

The paper gives my opinion of the state of the art of risk identification and assessment, discusses techniques which are used and gives examples and references to assessments in nuclear plant, chemical plant and transport, the handling of liquefied petroleum products and others. Some assessments are exploratory, others carried out in great depth.

The paper discusses the purpose of assessment, its uncertainties, its value, and pointers to further development.

1. INTRODUCTION

This paper gives my opinion of the state of the art of risk identification and assessment in specific areas of an industrial society and gives examples where assessments have been carried out; sometimes exploratory, sometimes at great depth. The report suggests some pointers for the future.

I have not attempted to give an exhaustive review, but to use material with which I am reasonably familiar; there will be many more and perhaps better examples than the ones I have used. Similarly I have not tried to write a mathematical treatise: complex mathematics is generally only a small part of risk assessment.

Risk assessment operates in a very sensitive area of our present society. It can and should play an important part in ensuring some balance of effort in reducing risk, but to do so it must be understood and wisely used.

2. WHAT IS THE PURPOSE OF THE ASSESSMENT?

In many situations it is known that some risk exists in most heavy industries, in the procurement and use of fuel, in civil engineering, transport and construction. The very high risks of injury and death during the last century have been drastically reduced, generally by learning from mistakes: 'engineering technologies have advanced not on the basis of their successes but on the basis of their failures' (Hinton 1957).

† Elected F.R.S. 19 March 1981.

Why should we not continue to follow the same pattern of evolution? It has worked in the past: design, construction and operation improved and accidents decreased. One answer is in the first report of the U.K. Major Hazards Committee: 'the pace of change associated with modern technology allows less opportunity for learning by trial and error. It is increasingly necessary to seek to get design and operating procedures right first time. Because of their present day size and throughput, there are now many plants throughout the world where a critical first mistake can result in disaster.'

Sometimes the assessments are quantitative to derive a numerical risk–consequence relation; others may be qualitative. Either may be used to establish a risk ranking order when alternative options are available. One example, now much studied, is the risks from different energy options, although there is no clear evidence that the observable random risks affect choice whereas the fear of, or uncertainty connected with, potential major hazards seemingly carries much weight. Additionally, the greatest risk may stem from phenomena little understood, such as the greenhouse effect, or indeed from the failure to accept some current risks and uncertainties: to do nothing may be dangerous.

The purpose of an assessment lies not only in the derivation of a numerical risk table or a ranking order, but, of perhaps greater importance, in the better understanding on the part of the assessor of the characteristics of the problem, the related phenomena, the uncertainties and the identification of areas worthy of research or further investigation.

Assessments are sometimes carried out as part of a public relations exercise to persuade potential opponents that a chosen route, process or site, is safe, or safer than alternatives, e.g. that nuclear power is safer than burning coal and that 'soft options' carry risk.

3. What precision is required?

A large range of industrial activities have occupational fatality rates within a factor of 4, from 25 to 100 deaths/10^6 million per year (U.K.). Hence if the exercise of assessment is to establish a ranking order in risk in industry, a factor of 2 might be important.

Is a factor of 2 always important? There are several aspects to this. If an individual risk (of death) is around 10^{-3} per year as derived for the Canvey Island Study (H.M.S.O. 1978), then a factor of 2 is important. If an assessed risk is very much lower (10^{-5} per year), a factor of 2 is barely discernible and would not carry conviction.

Similarly if a major hazard study gives a predicted consequence of 10 casualties, a factor of 2 is important; if the prediction is 1000, a factor of 2 is likely to lie with the range of uncertainty of the prediction and in general the order of magnitude is a sufficient pointer to the severity of the event.

There is no sense in trying to achieve greater accuracy in the consequence than can be achieved in estimating the probability of the event. It would seem inconsistent or illogical to follow one type of risk – or risk in one type of activity – to a degree of precision far higher than can be achieved with other competing or parallel risks, when they are of the same order of magnitude.

4. Over what period?

If the exercise is concerned with relatively short-term actions, today's data will apply. If the concern is with the future, as with an energy programme, the time of interest may be 20–50 years ahead or more. Today's data may not apply and some extrapolation of current risk rates will be necessary.

The time scale may be some thousands of years, as from nuclear power, coal burning, deforestation; it is then impossible to *prove* the rightness of any decision. However, a *no* decision or the absence of a decision may introduce grave risks over a short time.

5. What scale of risk or index of harm?

Several writers have explored this question (Pochin 1977; Reissland & Harries 1979). There is a difference between dying today and dying in 20 years, a difference between death or injury or later disability, and harm to this generation or the next.

One index of harm might be related to the loss of life expectancy, with a further weighting for injury. This type of index seems plausible but is seldom used. Many risk assessments add together deaths of all types, immediate and delayed and those derived from the summation of fractionally possible risks; e.g. a risk of 10^{-4} per year of an accident giving 10^4 casualties is sometimes equated to 1 per year. This loss of information, the smoothing out of different types of risk, obviously detracts from the value of the assessment.

Some form of index might be used to try out its value in various exercises, provided that it does not become enshrined and used without thought of its relevance in each case.

6. Siting

The early criteria, developed for the siting of nuclear power stations, were aimed at finding a relatively secluded site as a 'hole' in a population distribution. Some differences exist between criteria in the U.S.A. and in the U.K., probably arising from different population densities and distributions.

Early studies were deterministic, based on nominal releases of fission products. By 1967 a probabilistic approach was proposed (Farmer 1967), and in later papers, as by Beattie (1978), additional factors gain weight: global effects, contamination of land, protection of the public, and emergency procedures.

7. Residual uncertainties

What is meant by uncertainty? The U.S. Reactor Safety Study (Rasmussen 1975) assigned a multiplicative factor of 5 for uncertainty in its result.

When a safety study yields a wide range of consequences of varying severity over a range of probabilities, it is expected that the uncertainty will be greater for the less probable events.

A reactor safety study, or the like, should be carried out with reference to a specific design, with defined procedures for inspection, commissioning, operation and maintenance, and with the use of failure rate data most appropriate to the equipment of that design. The study conclusions would then be valid for that unit, but not necessarily for others. Unfortunately, Rasmussen (1975) has been quoted by many as though it applied to all reactors. An estimate of uncertainty should therefore take account of this, and no reasonable estimate should be given unless the second unit (reactor) is compared with the first for which the detailed analysis had been done.

Uncertainty is not constant with time. Confidence should improve with continued experience of assessment and of plant performance, and with improved coordination of inspection and maintenance in relation to the sensitivity of equipment in the fault tree analysis.

A study of common mode failures has been made by Edwards & Watson (1978) and Smith & Watson (1980), and uncertainty introduced by the human factor is reviewed by Embrey (1976).

8. Low-level exposure to risk

The risk to workers or the public from exposure to radiation is based on a fairly well established relation between exposure and effect, and an assumed linear relation at low levels of exposure. It is much more difficult to quantify the risks through exposure to chemicals; even if some toxicity data are available, they are unlikely to extend to low-level exposures, particularly as 'nutritional and other variables alter susceptibility to toxic effects by an order of magnitude or more – are poorly understood and little studied' (Royal Society 1978). However, some risk tables quote at a risk level of 1 in 10^6 the effect of smoking, drinking and eating various substances – necessarily assuming some relation at low exposure levels, no threshold and not taking into account the combined effect of different exposures.

9. Major hazards

A major hazard installation, or operation such as transport, is one that has the potential to hurt many people even if unlikely to do so.

Particular interest in major hazards has been engendered by such accidents as at Aberfan, Moorgate and the explosion at Flixborough. This was followed by an inquiry and the setting up of the Major Hazards Committee which has issued two reports through the Health and Safety Commission.

An accident on a major hazard site may result in only few or no casualties, as at Mississauga or Three Mile Island, although in either case, under other circumstances, there could have been tens or hundreds of deaths, immediate or delayed. The accident to the tanker carrying propylene at San Carlos (Spain) killed over

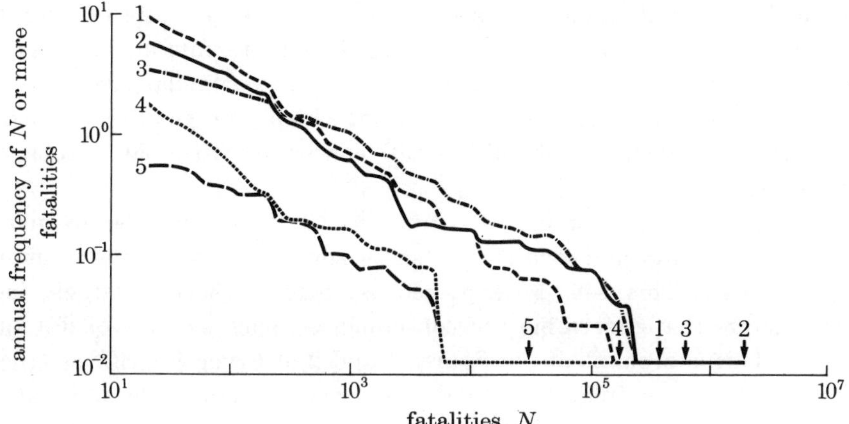

FIGURE 1. Frequency of multiple-fatality accidents occurring world-wide from natural causes. The numbered arrows indicate the endpoints of the curves. 1, Storms; 2, floods; 3, earthquakes; 4, avalanches and landslides; 5, volcanic eruptions.

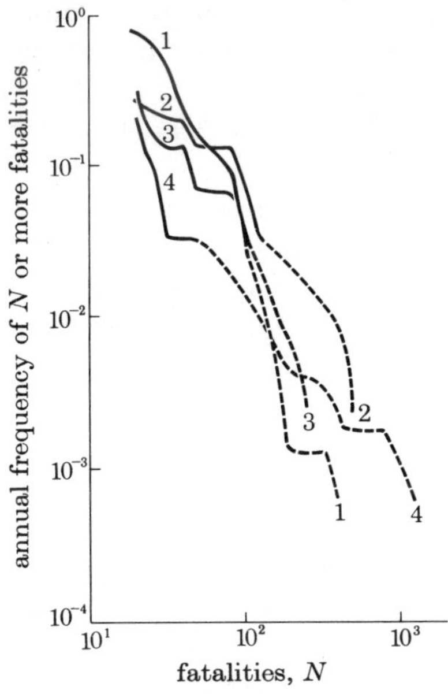

FIGURE 2. Extrapolation of British Isles accident data: solid lines, data on accidents occurring within the British Isles; broken lines, extrapolations based on world-wide data. 1, Aircraft accidents; 2, railway accidents; 3, mining accidents; 4, fires and explosions.

150 people; this would be an extremely severe result of a tanker accident, but not necessarily the worst or maximum, as it is always possible to invent a situation that is worse than any previously devised. It is then a question of probability.

Hence a description of a hazard requires a statement or estimate of probability and consequence. The hazard, in general, cannot be described as a single number, not maximum, mean nor likely. This is not at variance with the findings of Marshall (1977) and updated in the second report of the Major Hazard Committee (1979), in which a mortality index has been derived from a study of over 150 explosive accidents from liquids, gases or solids. This represents an 'average' of those that have occurred, whereas the identification of a particular hazard site or operation makes available more information from which a range of consequences may be deduced at varying probabilities, all consistent with past experience suitably adjusted.

It is to be expected for most hazard situations that there will be an inverse relation between frequency and severity, i.e. as the severity increases the estimated probability decreases. This is to be expected as there are several largely independent parameters that determine the consequences, such as time of day, and wind direction. This is brought out in figures 1 and 2 of Fryer & Griffiths (1979), presenting the relation of frequency to the severity of man-made and natural hazards.

Accidents from natural causes, floods, earthquakes, etc., on a worldwide basis may kill 10^5 to 10^6 people in one event, whereas man-made causes such as mining, fires and explosions are unlikely to kill more than 1000 people in one event, but this number could be increased by a factor of 10. However, the difference in the slopes of figures 1 and 2 may result from limitations on data. On figure 1, there would be few recorded data of events causing few fatalities (1–1000), whereas on figure 2 the time base is small and the extrapolation to 10^{-3} to 10^{-4} per year is weak.

As remarked in §5, it is hard to ascribe a weight to a predicted event causing, say, 10^4 casualties at an estimated probability of 10^{-4} per year; this is not equal to 1 per year, as seen by public reaction to multiple road accidents. Some writers have suggested that an apparent weighting might result from squaring the consequence, but I see little merit in this.

10. Some examples of risk assessment

(a) *Chemical plant*

In 1971 it was recognized that absolute safety could not be achieved in the design and operation of potentially hazardous chemical plant. A new plant was designed to a self-imposed risk target, and although the available data were limited at the design stage, they were improved by monitoring in later operation and related to maintenance and inspection periods. The risk was one of a gas reaction in a replacement plant and a 'failure rate' target was set at 5×10^{-5} per year (Stewart 1971).

Lawley (1974) reported that the progressive concentration of production into large single-train units and the increasing need to operate closer to risk situations require refined methods for eliminating problems at the design stage, and discussed operability studies and hazard analysis – a quantitative examination after a serious hazard has been identified.

In 1978, a major petrochemical site (Canvey Island) was assessed on behalf of the U.K. Health and Safety Commission after public concern about possible extensions. A quantified assessment was made of individual and societal risks of a complex site on which ten companies operate major installations. A major difficulty lay in obtaining and analysing relevant data and assessing the validity of the current interpretation of phenomena such as heavy gas flow. The flow of heavy gases is studied in theory and by experiment, an important problem in assessing risks and one not yet fully resolved (Kaiser & Walker 1978; Griffiths & Kaiser 1979).

Cremer & Warner (1978) set out a rational method for determining the appropriate layout and necessary safety zones between petrochemical plants and the community. It involves the quantification of both frequencies and consequences of possible hazardous incidents. The very nature of the materials handled in petrochemical plants presents hazards, and the presence of large quantities of these materials may constitute a major hazard.

(*b*) *Equipment*

There are many examples of reliability or safety related assessments of equipment: of pumps, valves, filters, fire detectors and ventilation equipment, and a massive collection of data for electrical, electronic equipment in the I.E.E.E. standard 500/1977.

For over 20 years there has been a special interest in the safety of pressure vessels for nuclear reactors. Between 1967 and 1974 there were extensive reviews of pressure vessel statistics in the U.K., Germany and the U.S.A.

Solomon *et al.* (1975) reported a simple methodology providing an interrelation between vessel weld integrity and inspection sample size, frequency and efficiency.

Marshall (1976) reported a study group assessment of the integrity of p.w.r. pressure vessels which considered material properties, quality control, inspection and the interpretation of crack growth on a probabilistic basis.

An important part of the pressure vessel analysis is the success rate in finding cracks by current inspection techniques. The results of an extensive European programme on the ultrasonic examination of three test plates give a challenging view of current inspection techniques (O.E.C.D./N.E.A. 1979).

There are many examples of quantified risk assessment of equipment and assemblies; one from France is applied to the design of a spent fuel cask handling system of a nuclear plant (Gachot & Barbet 1977).

(c) *Liquefied natural gas and petroleum gas*

There has been an upsurge of interest in the risk assessment of land and sea transport and the site of installations for liquid fuels. A U.S. Department of Environment report (1978) shows a wide divergence in various models of heavy gas dispersion. This distance to lower flammability limit varies from 5 to over 25 miles, although the consensus of the later studies narrowed this gap. Havens (1978) also discusses this problem.

The probability of hazard in transport is discussed by Jenssen (1978), Cave (1978) and Barlow & Lambert (1979), and transport by pipeline from St Fergus to Mossmorran and to Boddam, by the Health and Safety Executive (1978).

(d) *Nuclear waste*

The report by Inman *et al.* (1978) is exploratory, and concerned with the methodology of assessing risks associated with geological disposal of nuclear waste.

(e) *Chemical transport*

The potential hazard of transporting chlorine has been studied in the U.K., the U.S.A. and Finland. The U.K. study (Westbrook 1974) considered the relative risk of road and rail transport on a deterministic basis by assigning average numbers to the various parameters of accident frequency, severity, population density, etc., and reached a chance of one fatality every 100 years.

The transport in the U.S.A. was assessed on a probabilistic basis to maintain a distribution for the population and weather pattern and arrived at the chance of 100–1000 fatalities at 10^{-2} to 10^{-3} per year (Simmons *et al.* 1974). This is shown in figure 3 taken from the Rasmussen report.

A very similar study from Finland arrived at a lower risk commensurate with the lower volume of traffic (Lautkaski 1976).

(f) *Earthquakes*

Studies of earthquakes and effects on industrial plants have gained momentum through the need to design nuclear plant to survive acceleration of various types and severity. Wall (1979) gives a status report with many references. Hsieh & Okrent (1976) consider the seismic risk of nuclear reactors.

(g) *Aircraft and space*

In developing supersonic aircraft a safety target objective was proposed in the U.K. (Black 1968) to achieve a fatal accident rate of no more than 3×10^{-6} per flight or 1.5×10^{-6} per hour for an average flight of 2 h. It was recognized that there was a need to develop and use reliability engineering methods for the analysis of the safety of complex and interrelated systems. This method must lean heavily on probability analysis when this can genuinely be applied.

The study of defects in service and their rapid analysis and the subsequent actions have achieved a high safety record for the RB 211 aircraft engine (Thomas 1977).

(*h*) *Explosions*

To provide a background for the assessment of the risks from explosions, Marshall (1977) has collected and reviewed the records of past accidents, and similarly a U.S. review considered the range of effects of vapour-phase explosions near nuclear power plants (Eicher & Napadensky 1977).

Cerda & Napier (1977) have reported a hazard analysis in the context of sugar dust explosions.

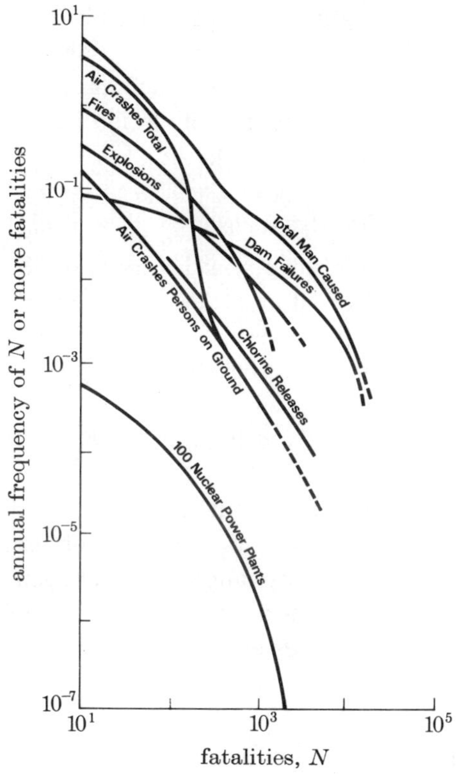

FIGURE 3. Frequency of man-caused events (fatalities due to car accidents are not shown because data are not available; car accidents cause about 5×10^4 fatalities per year).

(*j*) *General*

A widely based study of risk and risk–benefit of the various methodologies as applied to different technological systems has been reported by Okrent (1977). This includes specific applications for nuclear power.

The Delta Act of the Netherlands Government prescribes that their water defences should withstand specific probabilities that water levels may exceed, such as 10^{-4} per year for the 'heart of Holland' (Kleig 1977). A relatively new area for risk assessment lies in the application of genetic manipulation and particularly in its large-scale use (Brenner 1978).

(*k*) *Comparative energy risks*

There are many recent studies, not many of great merit particularly when scanning the whole subject.

These reports still leave open the questions of why, with what precision, for what period and with what projected result, and do not address the question of 'an index of risk'. There are many uncertainties, particularly the long-term effects, and also in low-level effects from effluents, particularly chemical. The illusion of safe, soft options is removed by these studies when all contributions to risk are included as by Inhaber. Although his early work is not accepted, it opened the door. There is often excessive effort to achieve accuracy, although swamped by uncertainties elsewhere (Health and Safety Commission 1978; U.N. Environmental Programme on Energy, 1979; Inhaber 1978).

(*l*) *Nuclear*

The improved assessment techniques now in use have been developed quite slowly over 20 years from the first effort (U.S. A.E.C. 1957), which assessed the consequences of the worst event (or nearly so), to the distribution of probability and consequences in the U.S.A., and the German light water reactor studies (Rasmussen 1975).

Some probability ideas were already being applied on safety limitation by the A.E.A. (Hughes & Horsley 1964).

Other papers, from 1957 to 1962, bring out the obvious link between reliability and safety and illustrate the gradual build-up of probability risk analysis to systems and situations in design, maintenance and operation.

There are many thousands of reports discussing the risks of nuclear operations, and there have been countless meetings and conferences.

11. Methods of assessment

Methods have been described in many publications; a good review is given by Okrent (1977).

Event trees and fault trees are in common use. These are systematic ways of proceeding from one event or fault to the sequential event or events that in turn lead to others.

Each event or fault may have several discrete consequences or a distribution of possible consequences.

Components such as pumps, valves, switches, etc., which form part of an installation to be assessed, may have different 'output states', i.e. they give rise to different consequences depending on varying combinations of 'input states', such as the availability of power, water or oil and the functioning of control and measuring devices.

The orderly presentation of this information may form a 'decision table', which then occupies a particular position in the fault tree construction.

Many analyses are quantitative. Discrete numbers or distributions are applied to probabilities and consequences of events and faults, so that the conclusion of the assessment is itself a range of possible consequences with assigned probabilities, with, if possible and relevant, an indication of the confidence level, or uncertainty, in the result. When some input data are clearly uncertain, even guessed, the effect of the guess should be followed through the analysis and indicated in the conclusion (as in the Canvey Island study a '*d*' symbol, p. 47). It is usual then to carry out a 'sensitivity analysis' to find the effect on the conclusion of varying the value of the input. This may be extended to a variety of input data to find the weak links in an engineered assembly or the assumed relevance of any phenomena.

Any risk assessment, or most, will require judgement throughout the construction of the trees and of course as to the value of the final answer. The orderly assembled judgement of a group of people may be called the 'Delphi approach'. This may be an iterative process with feedback leading to sequential revisions of judgement.

The basis of risk assessment is essentially simple; the task of carrying out an assessment may be difficult, but not many assessments require long event–fault trees or sophisticated mathematical techniques; they always require extreme care and skill in the perception and construction of the analysis and in the selection and use of data.

12. The value of safety assessment

It might be inferred, from the preceding discussion of uncertainty, that the main aim of an assessment was the production of a number or series of numbers or curves to represent the risk: often a frequency–severity plot. Many studies might start with this objective, but are then discredited if the results seem to be based on a poor foundation.

This was an early reaction to the U.S. Reactor Safety Study. The benefit lies in the study itself, such as the following needs.

1. To understand the system or plant in great detail. This may be part of a larger unit, such as a feed water system or auxiliary power system. This part of an exercise is often surprisingly difficult and may take 2–3 months.

2. To find out the likely performance and failures of the various plant items. Ten years ago there were only limited data, mostly electrical. The situation is better today, but is still developing.

3. To question the assumptions, particularly in relation to complex phenomena such as crack growth rate, aspects of depressurization, fluid flow and heat transfer, and light or heavy gas flow. This should combine with relevant research programmes.

4. To identify the weak points by formal analysis, event–fault trees or other techniques. Once those that contribute most to the assessed risk, or those that, if changed, would significantly change the risk, were identified, would be remedied by design, by improved inspection frequency, or by other means.

All are benefits if wisely used. It is an essential requirement that difficult exercises in the development of new technologies should be subject to constructive criticism, otherwise called a peer review. The open discussion of potential harm is difficult, particularly when, in the words of the Lewis report (1978), 'in the arena of reactor safety, a peer comment has come to mean anything written by anybody asserting anything about anything'.

Even with regard to this comment, an important benefit of risk assessment should be an improvement in communication, particularly and most usefully between designers, operators and licensing or approval organizations. It is better to say that the risk is 1 in 10^3 per year and agree or disagree about it, than to agree that the event is unlikely or hypothetical.

13. Who should do the assessment?

I cannot do better than quote from Kelly *et al.* (1979), as I fully agree with their comment:

> As the level of detail required by the reliability analyst increases, so do his demands on the designers' time and experience. At some point it becomes more effective to train the designer in reliability techniques than to train the reliability analyst in design techniques.

14. What further development?

Existing techniques are adequate for most safety studies. The main limitations lie in the relevance of existing data to the specific situation under review, and the limits of the assessor's perceptiveness in adequately covering all important sequences and in questioning the interpretation of phenomena. Improvement will come from continued application of risk analysis in many different fields, and from interaction with other groups involved in risk analysis.

In very complex situations, as for a nuclear reactor, some improvement in techniques will assist in making a more complete analysis. It would be useful if a tree study could take account of many different issues at any one junction; for example, valves, pumps, filters and fires are not adequately described by open/shut or yes/no, but by a continuous or multi-step function. Nevertheless, while it is tempting to seek a closer representation of a real plant, it is conceivable that the real gain may be small. The representation will still be that of a class of equipment and not that one which is in service.

There is room for improvement in data collection and classification for them to be meaningful in a reliability analysis.

15. Risk research

Reliability is closely associated with safety and the reduction of risk. In the U.K. the National Centre for Systems Reliability acts as a focal point to coordinate research projects in universities and technical institutions. This is one aspect of risk research. These programmes will include aspects of human factors, man–machine interfaces, mathematical modelling, computer code improvements and case studies.

What might be done to promote and improve risk analysis?

(*a*) Encourage risk or reliability analysis as part of an engineer's training.

(*b*) Identify a few 'technical' centres to carry out risk analysis.

(*c*) Form a link with these centres and with other recognized overseas units.

(*d*) Institute or encourage an annual meeting to review progress of techniques and case studies.

(*e*) Recognize that risk analysis, as with reliability, is not always glamorous and not always concerned with big risks; it has a substantial groundwork of small assessments, many of which are financially beneficial to the recipient.

(*f*) Initiate and support theoretical and experimental work investigating phenomena seen to be important in risk analysis, such as light or heavy gas flow or inspection techniques in difficult situations.

(*g*) Sponsor speculative investigations into other risks that might be related to the changing pattern of industry. Whereas the agglomeration and centralization of industry, energy production and distribution has led to the possibility of harm from major hazards, there may be risks to society through system instabilities, with the possible deprivation of resources.

I acknowledge considerable support from the Bundesminister für Forschung und Technologie in preparing this paper.

References (Farmer)

Barlow, R. E. & Lambert, H. E. 1979 *The effect of U.S. coast guard rules in reducing the probability of LNG tanker ship collision in Boston harbor*. Tera Corp.

Beattie, J. R. 1978 In *International Meeting on Nuclear Power Reactor Safety*, Brussels.

Black, H. C. 1968 *Aeronaut. Jl R. astr. Soc.* **72**, 115–122.

Brenner, S. 1978 *Nature, Lond.* **276**, 104–108.

Cave, L. 1978 *A comparison of the probabilities of spills of LNG and other liquid fuels in Boston harbor*. U.C.L.A.

Cerda, H. W. & Napier, D. H. 1977 In *Proc. National Conferences on Reliability*, Nottingham, p. 4.8.

Cremer and Warner 1978 *Guidelines for layout and safety zones in petrochemical developments*. C. 2056. London.

Department of Environment, U.S.A. 1978 *An approach to LNG safety and environmental control research*. EV 0002.

Dolphin, G. W. & Marley, W. G. 1969 *Risk evaluation in relation to the protection of the public in the event of accidents at nuclear installations*. U.K. A.E.A. AHSB(RP)R96.

Edwards, G. T. & Watson, I. A. 1978 *A study of common mode failures*, U.K. A.E.A. S.R.D. R.146.

Eicher, T. V. & Napadensky, H. S. 1977 *Accidental vapor phase explosions on transportation routes near nuclear power plant*. NUREG/CR0075.

Embrey, E. E. 1976 *Human reliability in complex systems: an overview*. N.C.S.R. R10. National Centre of Systems Reliability, Safety and Reliability Directorate.

Farmer, F. R. 1967 In *I.A.E.A. Symposium*, Vienna.

Fryer, L. S. & Griffiths, R. F. 1979 *Worldwide data on the incidence of multiple fatality accidents*. S.R.D. R149.

Gachot, B. & Barbett, J. F. 1977 In *Proc. National Conference on Reliability*, Nottingham, pp. 4.10. 1–23.

Green, A. E. 1977 In *Nuclear Systems Reliability and Risk Assessment Conference*. N.S.R. E.R.A. Gatlinburg, pp. 661–685. Philadelphia: Society for Industrial and Applied Mathematics.

Griffiths, R. F. & Kaiser, G. D. 1977 *The accidental release of anhydrous ammonia to the atmosphere. A systematic study of factors influencing cloud density and dispersion*. S.R.D. R154.

Havens, J. A. 1978 In *Fifth International Symposium on the Transport of Dangerous Goods by Sea and Inland Waterway*, Hamburg.

Health and Safety Commission 1978 *The hazards of conventional sources of energy*. London: H.M.S.O.

Health and Safety Executive 1978 *A safety evaluation of the proposed St. Fergus to Moss Moran natural gas – liquids and St. Fergus to Boddam gas pipeline*. London: H.M.S.O.

Hinton, Sir Christopher 1957 *The future for nuclear power*. Axel/Johnson Lecture, Stockholm.

H.M.S.O. 1978 *An investigation of potential hazards from operation in the Canvey Island and Thurrock area*.

Hsieh, T. & Okrent, D. 1976 *Some probabilistic aspects of the seismic risk of nuclear reactors*. U.C.L.A. Eng. 76113.

Hughes, H. A. & Horsley, R. M. 1964 In *B.N.E.S. Symp.*, London, pp. 198–203.

Inhaber, H. 1978 *Risk of energy production*. Atomic Energy Control Board publ. 1119, and revisions.

Inman, R. L., Helton, J. C. & Campbell, J. E. 1978 *Risk methodology for geologic disposal of radioactive waste. Sensitivity analysis techniques*. NUREG/CR 0394.

Jensen, T. K. 1978 Risk Analysis of LPG Transportation by Ship. In *Proc. National Centre of Systems Reliability R20*, pp. 185–202.

Kaiser, G. D. & Walker, B. C. 1978 *Atmos. Environ.* **12**, 2289–2300.

Kelly, A. P., Torri, A. & Emon, D. E. 1979 *The role of probabilistic analysis in the gas-cooled fast reactor programme*. GA A 15463.

Kleij, W. van D. 1977 In *10th International TNO Conference*, Rotterdam.

Lawley, H. G. 1974 *Chem. Engng Prog.* **70** (4), 45–56.

Lautkaski, R. & Mankamo, T. 1976 *Chlorine transportation risk assessment. Technical Research Centre, Finland, natn. Power Engng Lab. Rep.* no. 27. Helsinki.

Levine, S. & Veseley, W. E. 1977 In *Nuclear Systems Reliability and Risk Assessment Conference*, Gatlinburg.

Lewis, H. W. *et al.* 1978 *Risk Assessment Review Group Report to the U.S. N.R.C.* NUREG/CR 0400.

Major Hazards Committee 1979 *Second report of Advisory Committee on Major Hazards*. London: H.M.S.O.

Marshall, V. C. 1977 *Chem. Eng.* **323**, 573–577.

Marshall, W. 1976 *An assessment of the integrity of PWR pressure vessels*. U.K. A.E.A.

O.E.C.D./N.E.A. 1979 *Report from the Plate Inspection Steering Committee*. SEN/SIN (79) 13.

Okrent, D. 1977 *A general evaluation approach to risk–benefit for large technological systems and its application to nuclear power*. U.C.L.A. Eng. 7777.

Penland, J. R., Smith, A. M. & Goeser, D. K. 1977 In *Nuclear Systems Reliability and Risk*

Assessment Conference, Gatlinburg, pp. 176–731. Philadelphia: Society for Industrial and Applied Mathematics.

Pochin, E. 1977 *Problems involved in developing an index of harm.* I.C.R.P. 27.

Rasmussen, N. C. 1975 *An assessment of accident risks in U.S. commercial power plants.* WASH 1400.

Reisland, J. & Harries, V. 1979 *New Scient.*, 13 September, pp. 809–813.

The Royal Society 1978 *Long term toxic effects. A Study Group report.*

Simmons, J. A., Erdmann, R. C. & Naft, B. V. 1974 In *Natn Conf. on Control of Hazardous Material Spills*, San Francisco.

Smith, A. M. & Watson, I. A. 1980 In *Reliability Engineering*, p. 1.2.

Solomon, K. A., Okrent, D. & Kastenburg, W. E. 1975 *Pressure vessel integrity and weld inspection procedures.* U.C.L.A. Eng. 7486.

Stewart, R. M. 1971 In *Inst. Chem. Engrs Symposium Series* no. 34, pp. 99–104.

Thomas, F. H. 1977 In *National Conference on Reliability*, Nottingham.

U.N. Environmental Programme on Energy 1979 *Energy report series: The environmental impacts of production and use of energy*, pt 1 (*Fossil fuels*); pt 2 (*Nuclear energy*).

U.S. A.E.C. 1957 *Theoretical possibilities and consequence of major accidents in large nuclear power plants.* WASH 740.

Wall, I. B. 1979 In *Atomic Industrial Forum*, New York City. California: Nuclear Services Corporation.

Westbrook, G. W. 1974 In *Symposium of Loss Prevention and Safety.* Elsevier.

Discussion

D. J. Kuenen (*Blauwe Vogelweg 2a, Leiden, The Netherlands*). With regard to natural catastrophes, the Netherlands was mentioned in relation to floods. The risk was calculated to be 10^{-4}. This was based on the height of the sea-dams only in relation to the probabilities of high sea levels. This was a political decision rather than one arrived at by engineers or statisticians. (The risk level was estimated by the people of the Netherlands as the higher, the nearer they had been associated with the actual catastrophe in 1953, when 1835 people were drowned.)

D. Andrews (*Heathland, Cawston, Rugby, U.K.*). It may be convenient to use quantitative methods in the manner that Professor Farmer suggests, but before doing so it is necessary to establish that the practice is valid.

The man who designed the notorious pipe at Flixborough did not just guess its dimensions: he used a formula. This formula was well known and easy to understand and use. We know now that the formula used was wrong, because the pipe failed. We know also that we could have known that the formula was inappropriate because the circumstances of its application differed from the assumptions from which the formula was predicated. We have to make sure that we are not making a similar mistake ourselves with regard to quantitative assessments.

Probability theory, on which quantitative assessments rely, is a philosophical conception. That it is appropriate to use it for the purpose proposed is a hypothesis. There is a branch of philosophy that is devoted to the testing of hypotheses: it is known as epistemology. The epistemological rules that have been determined for establishing the validity of hypotheses may be summarized as follows: (i) the hypothesis must be general; (ii) the hypothesis must be consistent with all the facts; (iii) the hypothesis must be consistent with all other valid hypotheses; or, if it requires a change in those hypotheses, the amended hypotheses must still be valid; (iv) the implications of the hypothesis must also be valid.

A valid hypothesis may not be true, but an invalid hypothesis is false. If the use of quantitative assessments cannot be shown to be valid we are not entitled to rely upon them for protection against major dangers.

There can be no objection on epistemological grounds to reliability assessments where the population of causes is constant, the activation of any cause is brought about at random, and the number of cases is sufficient to give confidence in the statistics. In the quantitative assessment of the safety of any complete plant, however, there are two necessary types of factor about which one cannot have assurance that they satisfy these requirements, namely the factor allowed to cater for unanticipated unknowns, and the factor allowed to cater for the undependability of human performance.

In respect of unanticipated unknowns, one can only guess, and guesses have no epistemological validity.

In respect of human behaviour, the possibility of intelligent intervention in the selection of causes from the population of choice is specifically excluded in the archetype for which the fundamental postulates of probability theory were developed, and on which all subsequent probability theory depends. One is not allowed to mark the cards or peer in the urn. The population of causes is not constant or finite: people invent things, and have mental aberrations that extend the possibilities beyond the scope of mechanistic reliability. Moreover, there is evidence to show that human behaviour – even where the life of a person concerned may be at risk – may have no central tendency.

We are entitled to use probability theory in respect of human behaviour only where the preferences of any individual member of the sample can have no significant influence on the mean response of the whole population, and where an answer in terms of the mean is adequate for the purposes intended. In respect of the assurance of safety against major dangers, we are concerned with the response of the one particular person who has the power to influence the course of events, and probability theory has nothing to say about that.

Specifically, a probability estimation has validity only for the mean frequency of outcomes in an infinite repetition of events. A frequency distribution is not necessarily a probability distribution, and may not be so regarded unless the basic assumptions are satisfied. A probability prediction for an inhomogeneous population is true of no individual member of that population. And the population of causes to which any probability statistic refers is necessarily more restricted than the total population of uncertainty.

Taking all these aspects into account, it is evident that it is not safe to rely upon quantitative assessments for determination of what should be done to ensure safety in circumstances where any single accident is unacceptable.

The proposal to use quantitative assessments arises from a synthesis of two confusions: a confusion between value and validity, and a confusion between decisions on what has to be done and decisions on how to do what has to be done.

Only a deterministic assessment of what has to be done is valid, and the only available criteria for this are those of possibility and acceptability. Anything that is possible is credible, and all possibilities must be identified. Whether it is necessary to take specific steps to guard against any particular possibility depends on whether that particular outcome is acceptable. This, in turn, depends on whether the occurrence can be detected and steps taken to arrest and normalize the situation before unacceptable harm is done or before the hazard escalates.

Having decided that something has to be done, it is perfectly reasonable, and often desirable, to use probability theory to determine how best to do it, and for this purpose quantitative criteria may be set. For instance, the B.C.A.R. require the probability of a hazardous effect arising from an individual failure to be less than 10^{-8} per hour of engine operation. Nevertheless, in many cases it is just not practicable to determine these probabilities, and reliance must be placed on the effectiveness of imposed controls.

Probability theory must not be allowed to become a scientific superstition. It is not an incantation for the exorcism of unthinkable events.

There is need for a forum in which the validity of proposed safety philosophies can be publicly debated.

F. R. Farmer. We have previously discussed the merits or shortcomings of probabilistic risk analysis. It can be extremely useful but should never be the sole input to a decision.

A probabilistic analysis can identify that which is possible just as effectively as a deterministic study. The examples in my paper concerned with the transport of chlorine show this. The deterministic study of Westbrook deduced the 'possibility' of one death in 100 years. The probabilistic studies of Simmons and of Lautkaski showed the 'possibility' of killing 100 or 1000 people, but also indicated an assessed probability. I prefer these latter two studies as they retain more variables in the analysis and provide a wider base on which to make a decision.

Proc. R. Soc. Lond. A **376**, 121–131 (1981)
Printed in Great Britain

The risk of producing energy

By H. Inhaber
Oak Ridge National Laboratory, Bldg 2001, Oak Ridge, Tennessee 37830, U.S.A.

If risk is to be understood by the public, the sometimes abstruse numbers and calculation must be put into perspective. In terms of energy, the simplest way is to compare the risk per unit output for different forms of production. Previous studies have suggested that coal- and oil-fired electricity have higher overall risk than nuclear power, due primarily to the former's air pollution effects. While newer, non-conventional energy forms such as solar and wind power may appear at first glance to be risk-free, this is not so. Because of the diluteness of the incoming energy, these non-conventional systems require large numbers of collectors per unit energy output. In turn, considerable quantities of steel, glass, copper, aluminium etc., are required, the production of which incurs occupational risk. In consequence, non-conventional systems can have substantial risk to health. The implications of this and Siddall's recent calculations on cost–benefit analysis are explored in terms of public policy.

Centuries ago, William Shakespeare wrote, in *King Henry IV*, 'Out of this nettle, danger, we pluck this flower, safety.' If there is one area where the public seems interested in plucking blossoms of safety, it lies in the field of energy production. Claims and counter-claims have been made with increasing vigour about which energy systems are safest for human consumption, so to speak.

This paper will consider (*a*) how the risk of any energy system can be calculated, at least crudely, and (*b*) the relation of nuclear risk to costs, as discussed in a recent paper by Siddall (1980).

Dangers in energy production

Risk to human health has become one of the prime considerations in choosing energy systems, if it has not already overtaken economic cost. Accidents at reactors, collapses of dams – these are what capture public attention. But are they the total risk of energy systems? Just as the cost of any article is made up of a number of sources, so the risk of an energy system has many origins. While it is impossible to state with certainty that the sources of risk noted in figure 1 are the only sources, they are the major considerations in most risk studies.

Most of the risk sources in this figure seem to be of non-catastrophic origin, i.e. accidents or illnesses that occur one at a time. This is confirmed in the report from which figure 1 is taken (Inhaber 1978). Even for systems like nuclear power and hydroelectricity, which receive considerable publicity about real or potential catastrophes, the proportion of catastrophic (as measured by historical statistics) to non-catastrophic risk is very small.

It might be contended that catastrophic risk is the only aspect that matters, and that all other risk should be neglected in calculations. This is a philosophical point, and it cannot be proved one way or the other. However, it seems clear that the total loss to society from building and operating energy systems is made up of all the deaths, accidents and illnesses that can be attributed to them, not just some. In the words of Oscar Wilde, we should add up all 'the sordid perils of actual existence'.

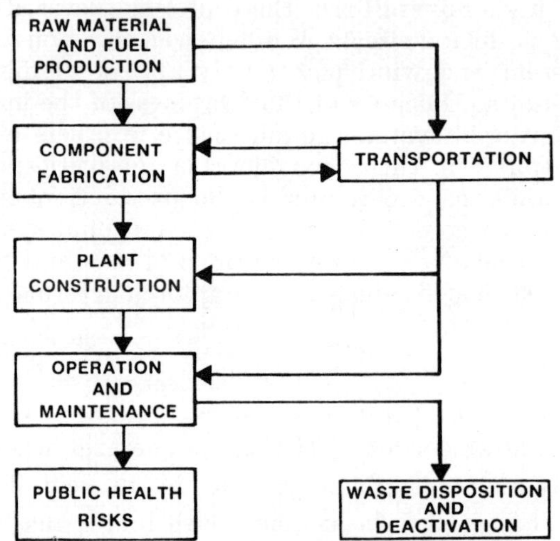

FIGURE 1. Sources of risk in energy production. The relative importance of each component depends on the energy system. For example, there will be no fuel requirement for most non-conventional (not in widespread use) systems, although all require raw materials. Public health risk due to coal is primarily from air pollution, and that due to nuclear from waste management and possible reactor accidents. Transportation plays an important part in producing energy.

Does the production of energy actually increase occupational risk (as contrasted with public risk), or would this risk have existed in any case? To illustrate this seemingly obscure point, consider smelting steel for a turbine destined for a steam-fired electrical plant. It is agreed that producing a tonne of steel has associated with it a number of deaths and accidents. If the steel had not gone into the turbines, it might have been incorporated into lorries or tricycles. In this sense, the overall risk to society has not increased. By this reasoning, only if more steel than normal is required does the risk rise.

This approach, sometimes termed 'incremental', has not been followed in most risk studies (see, for example, Inhaber 1978, 1979; Comar & Sagan 1976; Hamilton (ed.) 1974; Gotchy 1977) for a variety of reasons. First, no account is taken of public risk in the 'incremental' reasoning. As noted above, public risk of energy systems has usually drawn more attention than occupational risk. Secondly, the cost of anything, whether in terms of risk or money, is made up of what does take place, not what might have taken place. For example, consider an automobile. The steel, copper, glass, aluminium, etc., used to build it might well have been employed elsewhere, but its cost to the purchaser is based on the materials actually used. In the same way, almost all risk analysts use the 'absolute' approach, in which the total risk is computed, rather than the incremental.

Figure 1 indicates that most of the risk of any energy system will be 'statistical' rather than direct. Direct risk can be defined as that intimately connected with building and operating a system: falling off the roof while installing a solar collector, dying in a coal mine accident, and so forth. But there will be also substantial risk from transporting copper for wiring, producing Fibreglass for windmill blades, and all the myriad activities that result in a complete energy system. This latter risk is indirect or 'statistical'.

One can make a case that 'statistical' risk is not part of total risk, since it is not directly associated with the energy system. On the other hand, a considerable portion, at least in terms of attention paid, of the risk of nuclear power and fossil-fuelled systems is statistical in nature. For example, it has been estimated (*Ad Hoc* Population Dose Assessment Group 1979) that there will be about one death as a result of the Three Mile Island nuclear accident in 1979. This cancer death will occur over the next few decades, in an area where hundreds of thousands of cancer deaths will take place in the same period. Because this death cannot be identified explicitly, it is clearly 'statistical'. The same principle applies to many of the deaths and illnesses caused by air pollution.

Few, if any, have suggested that because these deaths are 'statistical', they should be discarded from risk calculations. If the concept can be used for nuclear power and fossil fuels, there seems to be no reason why it should not be used to compute the risk of solar, wind-power, or any other energy system.

Calculating risk

How does risk calculation proceed? As can be deduced from figure 1, there are a variety of methods used. The scope of this paper does not allow a full discussion of each type, but the risk attributable to material acquisition is illustrated in figure 2, which shows, in the words of the Earl of Chesterfield, a 'chapter of accidents'.

The amount of materials required for each system per unit energy output over its lifetime must be determined. The number of man-hours (or person-hours) required to produce this quantity is then estimated. Labour statistics that show the number of deaths, accidents and illnesses per man-hour are generally available.

The three sets of data are then combined for each material. For example, suppose mining X tonnes of coal requires Y man-hours. If the number of deaths per man-hour of work is Z, then the number of deaths per tonne of coal is YZ/X.

The risk associated with each material is found in the same way, and these sub-values are added to determine the total risk of material acquisition.

Finished products will require intermediate and raw materials, and this should be accounted for in the calculations. For example, steel will require coal, iron ore and other industrial goods.

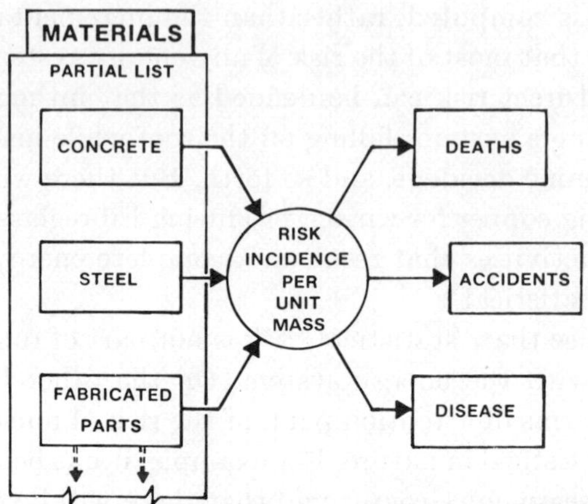

FIGURE 2. Risk from material acquisition. Each of the raw materials in an energy system has an associated risk. The risk depends on the accident, illness and death rate per unit mass produced in the appropriate industry. Dashes indicate that other materials are used.

A similar approach may be taken in calculating construction risk. The trades – electrical work, plumbing, roofing, etc. – must be determined, as well as the time required per unit energy output. Then the risk per unit time (as contrasted with unit mass) can be found from industrial statistics. The data are then combined as above, and added to the material acquisition risk.

The other components of risk shown in figure 1 are calculated in similar ways when possible, although different techniques are sometimes required. For example, transportation risk is calculated by assuming that all materials move a fixed distance by rail, and the average risk per tonne-kilometre is used to evaluate this source. The main exception is that sand and gravel are assumed to move negligible distances from where they are quarried.

A point of possible contention is how one calculates public risk for rare events like dam failures and reactor radioactivity releases. There are basically two approaches: historical and theoretical. The historical method takes what has happened over time as the basis for computations. Since there have been no theoretical calculations for the frequency of hydroelectric dam failures, the number of people killed by these accidents is divided by the total amount of hydroelectricity ever produced to yield public risk from this source.

On the other hand, nuclear power has had at least two extensive theoretical studies of public risk (N.R.C. 1975; Federal Minister of Research and Technology

1979). Values based on these studies, taken from a number of literature sources, were used by Inhaber (1978, 1979). In general, 'maximization' of nuclear risk, i.e. taking the highest value from these sources, was used, to avoid potential charges of pro-nuclear bias. This procedure was not followed for the other ten systems considered by Inhaber (1978, 1979).

One may contrast historical with theoretical estimates of nuclear public risk. Generally speaking, the historical (or experimental) values of risk are lower than the theoretical ones, even when the accident at Three Mile Island is taken into account. Of course, there is no guarantee that this situation will remain unchanged in the future, but it is of interest to note that the controversial theoretical studies (N.R.C. 1975; Federal Minister of Research and Technology 1979) have not underestimated nuclear risk, at least so far.

'Worst cases' and experimental data

As noted above, nuclear risk was 'maximized'. This does not mean that the worst case imaginable was used. One can easily conceive of 'worst cases' for any energy system, in which the fraction called risk per unit energy is very large. Examples that come to mind are: Italian hydroelectricity in 1963, the year of the Vajont disaster, the largest in history; ocean thermal gradient systems off Africa in the 1950s, which produced very little energy; or Pennsylvania nuclear power production in 1979, the year of Three Mile Island. 'Worst cases' were not used for any system considered by Inhaber (1978, 1979), and have generally not been employed in other comparative risk studies.

While most data in risk studies are fairly straightforward, a distinction should be made between present-day experimental data and those of the future. The distinction can be important in terms of certain non-conventional systems, since some of their advocates have suggested that the efficiencies of these systems might improve strongly in the future. This would produce more energy per unit materials and labour used in their construction, and so the risk per unit energy output would decrease. For example, photovoltaic systems have, at present, an efficiency (ratio of energy output to incoming solar energy) of 5–10 %; it has been stated that this may rise to 20 % in the future. Windmills usually have a load factor (ratio of energy produced to maximum possible production) of 15–20 %, but some believe that this could be much higher in coming years. Regardless of these beliefs, it seems reasonable to use current data to construct risk analyses, since the future rarely comes about the way we expect.

Results

Given all these general considerations, and others not mentioned here owing to lack of space, what are the results of risk calculations? Some findings, from using the methodology briefly sketched above, are shown in figure 3.

What accounts for the relative rankings? Simply put, most non-conventional systems – on the right of figure 3 – employ dilute energy like sunlight or wind. This energy is dilute in contrast to the concentration in a lump of coal or a nuclear fuel rod. It then requires a large number of collectors, windmills, solar panels, etc. per unit of useful energy. In turn, this array of collectors requires considerable amounts of steel, glass, copper and other materials. Producing the raw, intermediate and finished materials, fabricating and installing them generates industrial risk. This recitation may sound rather like 'the house that Jack built', but it suggests the origin of figure 3.

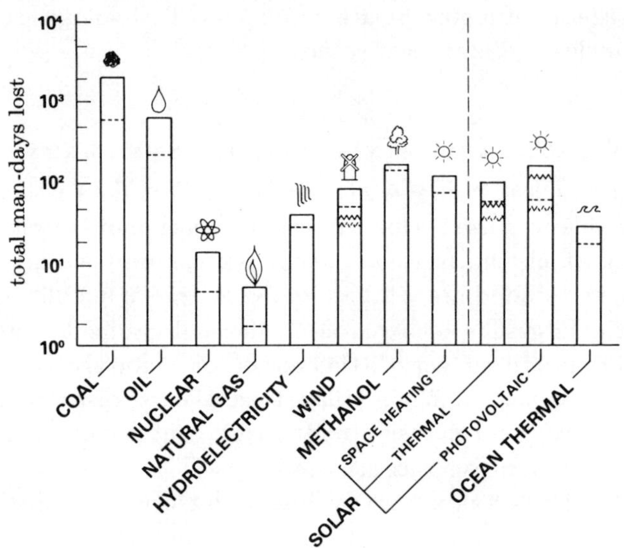

FIGURE 3. Total man-days lost per megawatt-year net output. Public and occupational deaths are combined. The top of the bars and the dotted lines indicate the upper and lower part of the range, respectively. Bars to the right of the vertical dotted lines indicate technologies less applicable to Canada. Note the logarithmic scale. Deaths are assumed to contribute 6000 man-days lost per fatality, and are combined with accidents and illness on that basis. Solid and dashed jagged lines on the three of the bars indicate maximum and minimum values, respectively, when no back-up (or low-risk back-up) energy is assumed.

On the other hand, relatively concentrated sources like coal and oil also have high risk; in fact, the highest of all. This is primarily due to the air pollution effects from these systems, although the risk of coal mining also plays a part.

Some non-conventional systems may also require back-up in the form of conventional energy for the times when there is no sunshine or wind. In figure 3, results are shown with and without back-up for the three systems assumed to require it. While the absolute values of risk clearly decrease when no back-up (or low-risk back-up, such as natural gas or nuclear) is assumed, the relative ranks remain about the same. For example, solar photovoltaic drops from third to fourth, based on maximum values; wind remains the same, at seventh.

NUCLEAR RISK

Because of the concern often expressed about nuclear risk, some words about it are in order, although not much detail can be given in a few paragraphs. The reasoning behind the choice of values for potential catastrophic reactor releases was given above. A comparison of nuclear with hydroelectric disaster risk suggests that the latter, based on actual events, has a risk of 2.5 to 13 times the former, depending on whether deaths or total man-days lost are considered.

Risk due to nuclear wastes was generally not evaluated until the last few years, since it was assumed that plutonium would be reprocessed. Recently, estimates of the risk due to wastes have been made (A.E.C.; Erdmann *et al.* 1979; Karn-Bransle-Sakerhet 1978). While they differ in their assumptions and results, all quoted here show a very low risk per unit energy output from the nuclear system. Of those used by Inhaber (1978, 1979), the man-days lost per megawatt-year ranged from 0.03 to 1.5. It may be seen from figure 3 that this is small. Of course, until wastes are actually disposed of, there will be some uncertainty in these values.

Figure 3 shows that nuclear power ranks second lowest in total risk, somewhat above electricity produced from natural gas. The two systems would have been closer if nuclear risk had not been maximized, for the reason given above.

Nuclear occupational risk is relatively low, because it requires considerably less mining than coal. Although its potential public risk has received considerable attention, the studies quoted suggest that this quantity is also low. The combination of the two components implies low total risk.

NUCLEAR RISK AND COSTS

Because Inhaber (1978, 1979), the work on which much of the above text is based, is a Canadian study, another recent Canadian paper (Siddall 1980) on the implications of nuclear risk deserves some mention. The questions that it raises are, how much can society spend to reduce risk, and are we increasing risk in one place by reducing it somewhere else?

Consider an example. Risk to the public from reactors is non-zero, although probably small. In principle, this risk could be reduced even further by building more concrete containment systems than the one per reactor used in the West, or even building reactors underground. On the other hand, extra occupational risk would be produced by the new types of construction. It is by no means clear that the total risk, made up of public and occupational categories, would be reduced by this proposal.

Siddall (1980) makes similar points in his paper. Noting that the estimated cost of saving a life varies strongly from one activity to another, he states that the cost of saving lives in nuclear power generation is high, if the theoretical and experimental data on safety is to be believed. He notes that while it may sound cruel and heartless to suggest it, transferring some money now spent on nuclear safety to other life-saving activities may actually reduce the overall risk to society. At present, however, there are few who would make this argument in public, regardless of its merits.

The conclusions of Siddall (1980) are in broad agreement with those deducible from figure 3, which shows that all energy systems produce some risk, and those that are sometimes assumed to be risk-free can have substantial risk when the entire energy cycle is considered. Clearly, more work needs to be done to ensure that society does not increase risk in the process of trying to reduce it.

Summary

The study of risk has become more important as we face more difficult choices with respect to energy. As research proceeds on non-conventional energy systems, it will be of interest to see whether as much consideration is given to reducing their risk as has been given to some conventional systems.

While some of the statements made in recent years about energy risk have undoubtedly been overblown and hysterical, society's attention has in the process been drawn to a sometimes neglected aspect of energy. In the words of Francis Quarles, 'Let the fear of a danger be a spur to prevent it; he that fears not, gives advantage to the danger.'

References (Inhaber)

Ad Hoc Population Dose Assessment Group 1979 *Population dose and health impact of the accident at Three Mile Island Nuclear Station.* Rockville, Maryland: Bureau of Radiological Health, U.S. Department of Health, Education and Welfare.

A.E.C. *High-level radioactive waste management alternatives.* Report no. WASH–1294. Washington, D.C.: U.S. Atomic Energy Commission.

Comar, C. & Sagan, L. 1976 *A. Rev. Energy* **1**, 581.

Erdmann, R. C. *et al.* 1979 *Status report on the EPRI fuel cycle accident risk assessment* (Report no. NP–1128), pp. 1–4. Palo Alto, California: Science Applications Inc.

Federal Minister of Research and Technology 1979 *The German Risk Study: summary.* Cologne: Gesellschaft für Reaktorsicherheit.

Gotchy, R. L. 1977 *Health effects attributable to coal and nuclear fuel cycle alternatives.* Draft report no. NUREG–0332. Washington, D.C.: U.S. Nuclear Regulatory Commission.

Hamilton, L. D. (ed.) 1974 *The health and environmental effects of electricity generation: a preliminary report.* Upton, N.Y.: Brookhaven National Laboratory.

Inhaber, H. 1978 *Risk of energy production.* Ottawa: Atomic Energy Control Board.

Inhaber, H. 1979 *Science, N.Y.* **203**, 718–23.

Karn-Bransle-Sakerhet 1978 *Handling of spent nuclear fuel and final storage of vitrified high level reprocessing waste,* vol. 1, pp. 115–119. Solna, Sweden: AB Teleplan.

N.R.C. 1975 *Reactor safety study: an assessment of accident risks in U.S. commercial nuclear power plants.* Report no. WASH–1400. Washington, D.C.: U.S. Nuclear Regulatory Commission.

Siddall, E. 1980 *Nucl. Safety* **21**(4), 451.

Discussion

H. J. Dunster (*Health and Safety Executive, London, U.K.*). I have some doubts even on the occupational deaths. I suggest that there are difficulties in attributing death 'upstream'. If a man were not at risk making steel for windmills he would be at some other risk. We need some concept comparable to the 'opportunity cost' in economics.

H. Inhaber. Mr Dunster has raised an interesting point, noting that workers in an energy cycle could, in principle, be employed somewhere else. Therefore, it might be difficult to attribute their associated risk to a particular occupation in an energy cycle. The point is a philosophical one and not easily answered. However, workers who are producing windmills or mining coal are engaged in those very tasks, and it is only speculation to guess whether their risk would be higher or lower if they worked at other trades. Leaving aside non-conventional systems like wind for the moment, every comprehensive study that I have read of conventional systems like coal and nuclear power has attempted to include coal and uranium mining risks without hypothesizing about what risk that these miners would incur if they left their jobs. This suggests that the approach I have followed, if not satisfactory to all, is reasonably standard.

To use an economic analogy along the lines mentioned by Mr Dunster, consider the price of a car. The steel, plastic, copper, cloth, etc., could have ended up in a television set, but this does not make the parts in a car free. The same reasoning applies to risk.

G. Van Reijen (*Commission of the European Communities, Brussels, Belgium*). It is clear from Dr Inhaber's figures that risks of electricity production from different sources, expressed in risk figures per megawatt-year, do not differ more than a factor of 10 for occupational risk and not more than a factor of 100 for public risk. As the error rate of these calculated risks is about a factor of 10 in both directions, it is difficult to conclude, from these figures, which of the forms of electricity production is the safest and should be chosen. The positive conclusion may be that none of these production forms is much more dangerous than the others.

The only exception to this is natural gas; for this source, the risk lies below the ranges mentioned. It seems that, from the point of view of risk, natural gas is an ideal energy source, but the available quantity is restricted and this circumstance sets a limit in its use. In my opinion, it would be of interest to compare for natural gas the two cycles: direct use for room heating and industrial purposes, and indirect use via electricity. In such a comparison, however, the risk factor may play an insignificant role.

H. Inhaber. I am afraid that I cannot agree that there is an inherent error rate of a factor of 10 in both directions in the calculated risks. Without seeing Dr Van

Reijen's detailed arithmetic, I am not able to comment fully on his statement. However, my analysis indicates that while there is some uncertainty about lifetimes, models of energy systems, etc., the uncertainty is much less than Dr Van Reijen states. For some aspects of risk, such as the number of accidents per man-hour in many occupations, the uncertainty is fairly small, since the data are tabulated carefully.

Dr Van Reijen goes on to discuss the effect of uncertainties on relative risk ranking. Without going into too much detail, I would say that most of the non-conventional energy systems were given the benefit of the doubt with respect to risk, i.e. longer lifetimes than have been achieved to date were assumed, a relatively high value of energy collected per unit area was specified, etc. As a result, I would suggest that the relative risk of most non-conventional systems that I considered may be even higher in comparison with natural gas and nuclear than shown in the graph, rather than similar.

The point about natural gas is well taken. The efficiency of burning natural gas in the home is claimed to be 60–70 %, in comparison with the efficiency of 30–35 % achieved in producing electricity. This greater efficiency would substantially reduce the risk associated with the total energy cycle. On the other hand, burning gas in the home may introduce certain air pollutants, the health effect of which is not known at present. Certainly, more study is required.

A. Robson (*C.E.G.B., London, U.K.*). The analysis by Dr Inhaber incorporates, in several places, estimates of the risk due to fossil fuel air pollution that derive from epidemiological studies by Lave & Seskin in the U.S.A. This probably also applies to the Health and Safety Executive study.

I consider that this work suffers from severe limitations, such as the assumption of a no-threshold linear dose–response relation, which make the derived numbers meaningless at the sort of exposure levels contributed by power stations in the U.K. The use of such uncertain data in comparative studies can thus lead to misleading conclusions regarding relative degrees of risk. The epidemiological approach also suffers from difficulties in relating pollution measurements to actual population exposure. This is because of the limited number of pollution measuring sites, population mobility (daily or long-term), exposure in the home and at work, previous exposure history and, as has been highlighted earlier, the significant and confounding effects of smoking. At the same time it is difficult to assess the impact on health of various climatic and socio-economic factors (J. Pickles & M. V. Proctor, Report no. RD/L/N10/80, C.E.G.B. (1980)).

Air pollution levels have improved significantly in the U.K. in the last few decades and generally compare well with air quality guidelines set by the World Health Organization, the European Community and others. Thus, while risk analysis can provide a useful overall view and in some cases define areas for further research, the numerical values produced have to be used with great care, taking into account the uncertainties involved. We must not lose sight of the levels of safety being achieved in absolute terms and the need to balance cost against the degree of benefit that can be obtained by improving levels even further.

H. Inhaber. Mr Robson raises the issue of the health effects of air pollutants like sulphur oxides and whether a threshold value for these pollutants is justified. There seems to be a difference of opinion with respect to this question on the two sides of the Atlantic, with scientists and governments in North America apparently adopting the no-threshold approach, and European authorities holding to a contrary opinion. I am not personally in a position to judge between the two approaches, although I admit that I used the no-threshold estimates in my calculations.

My calculations are easily modified if one wishes to assume a threshold for sulphur oxides. Lack of space precludes anything but a mention of the final results, but I find that the non-conventional energy systems to the right of figure 3 then have the highest risk values, interchanging their position with coal and oil. The latter two become intermediate between the non-conventional systems and natural gas and nuclear power. The absolute level of risk decreases for all systems, although by far the greatest decrease is for coal and oil.

Proc. R. Soc. Lond. A **376**, 133–149 (1981)
Printed in Great Britain

Industrial risks

By D. Okrent
Department of Chemical, Nuclear and Thermal Engineering, University of California, 5532 Boelter Hall, Los Angeles, California 90024, U.S.A.

A brief look is taken at the status of progress, or the lack of it, towards a quantitative approach to the estimation and assessment of risk for several technologies in the United States. The increase of interest in the consideration of comparative risks in decision-making is also discussed. Finally, a recently proposed trial approach to quantitative safety goals for light-water nuclear power reactors is summarized. The approach is divided into two major tasks: the predominantly social and political task of setting safety criteria, and the technical task of estimating the risks and deciding whether the safety criteria have been met. The safety criteria include the following: limits on hazard states within the reactor; limits on risk to the individual; limits on societal risk; a cost-effectiveness criterion as low as reasonably achievable; a small element of risk aversion.

1. Introduction

Industrial risk is a very broad topic covering very many different technologies and types of hazards. In principle, it includes both occupational risk and risk to the public health and safety.

In this paper I shall briefly discuss the current status of the safety approach being taken in the United States to three technological risks: dams, the storage of hazardous chemicals, and the disposal of hazardous chemical wastes. I shall then summarize a preliminary approach to quantitative safety goals for light-water nuclear power reactors, which was very recently forwarded to the U.S. Nuclear Regulatory Commission by its Advisory Committee on Reactor Safeguards.

2. Dam safety in the U.S.A.

According to a 1975 U.S. Army Corps of Engineers inventory there are about 50000 public and private dams in the U.S. that are 25 feet (*ca.* 7.6 m) or more in height and have a maximum impounding capacity of 50 acre-feet (*ca.* 6.2 ha m) or more. About 89 % of the dams are non-federal and about 40 % present a significant or large hazard potential to downstream life, if they should fail.

More than 100 large dams in the United States have failed since 1930. Buffalo Creek and Canyon Lake Dams are two well known dam failures from a long list that begins with the South Fork Dam near Johnstown, Pennsylvania in 1889. Other failures include the Saint Francis Dam in 1928 and the Baldwin Hills

Reservoir in 1963, both in California, the Teton Dam in Idaho in 1976, and the Barnes Lake Dam near Toccoa, Georgia in 1977 (Comptroller General 1979).

Dam disasters in the early 1970s caused about 355 deaths and extensive property damage. This situation led to the passing of the 1972 National Dam Inspection Act, which directed the Corps of Engineers to inspect most of the nation's dams and to recommend a comprehensive national programme for dam safety. By November 1976, however, no money had been appropriated and no inspections had been conducted. The Teton Dam failed in June 1976, prompting a number of Government actions including the development of guidelines for federal dams and appropriation of a modest amount of money for beginning the inspection of non-federal dams. In December 1977, after the failure of the Barnes Lake Dam, which killed 39 people, a higher priority and greater resources were devoted to the effort. A considerable number of non-federal dams have since been termed unsafe by the Corps of Engineers, but, for the most part, little action has been taken by the states in which they are located, partly because of a lack of resources.

During the San Fernando earthquake in 1971, the hydraulically filled, earthen Van Norman dam in Los Angeles was subject to gross liquefaction and was nearly breached. It was half full. Had it been full, experts believe it would have failed suddenly (Seed & Lee 1973), with the possibility of drowning up to about 100000 people living in its inundation plain (Ayyaswamy *et al.* 1974). The State of California passed a dam safety law soon thereafter and has instituted fairly stringent requirements on all non-federal dams within its confines. The Division of Dam Safety must make a finding that a dam is safe; however, it has not stated what quantitative criteria, if any, it uses in deciding on a determination regarding safety. To the best of my knowledge, neither the U.S. Corps of Engineers nor other federal agencies have quantitative risk acceptance criteria for their dams, although there is currently an effort to try to use such criteria in deciding on acceptable levels at the Jackson Lake dam, which is operated by the U.S. Water and Power Resources Service (U.S. Bureau of Reclamation 1978). Also, to my knowledge, the Corps of Engineers has not to this date performed any probabilistic analyses on the safety of their own dams.

In my opinion, dam safety provides an excellent illustration of how society reacts to large accidents. Dam failures have goaded the federal government and the State of California into action to improve safety for previously existing hazards. But there has not been a requirement for failure frequencies that would be representative of the high degree of risk aversion to large consequence events that some others have derived from an examination of historical events (Slesin & Ferreira 1976).

3. Risk to the public from chemical plant

Except for a few, recent proposed facilities for the importation of liquified natural gas (l.n.g.), there has not been published in the United States a risk analysis of a chemical installation similar to that performed for the Health and

Safety Executive (1978) in the United Kingdom for Canvey Island. Nor does the U.S. have a regulatory approach for acute hazards to the public from chemical installations similar to that being developed in the U.K. L.n.g. facilities are subject to stringent siting regulations in the State of California. However, in California and elsewhere, other chemical plants have generally been subject only to relatively modest regulation by local fire departments and building safety departments. Some chemical companies have taken relatively costly steps to improve the safety of their installations, or have even given up specific plants where the potential accident liability was large compared with the net profit. Such matters are kept company-confidential or proprietary, and little has been published in this regard. To the extent that I am able to ascertain the safety design criteria that are being used in such an upgrading of safety, I do not believe that they are nearly as stringent for chemical storage facilities within striking distance of large population groups and containing very large quantities of hazardous material such as chlorine, ammonia and l.p.g., as the U.S. Nuclear Regulatory Commission imposes on nuclear reactors. And for many, if not most, companies, no special safety measures beyond the relatively modest requirements of local government have been imposed.

4. Disposal of hazardous chemical waste

The matter of disposal of hazardous chemical waste, both as it represents a potential local source of hazardous emissions and as a pollutant of drinking water, is rapidly becoming one of the major public issues in the United States.

What is the magnitude of the hazardous waste that is disposed of annually? The U.S. Environmental Protection Agency (E.P.A.) estimates that approximately 57 Mt of the 380 Mt of liquid and solid industrial waste generated in the U.S. in 1978 were hazardous wastes, and that about 80% of these hazardous wastes are improperly disposed of in land fills or lagoons and pose a threat of groundwater contamination (E.P.A. 1980*a*). Very few land disposal sites are lined, and few have leachate collection systems. From their surface impoundment assessment, E.P.A. estimated that 5000–6000 industrial impoundments may contain hazardous wastes; few have a liner, and virtually no monitoring of groundwater is conducted to detect contamination beneath the sites; almost one-third have a high potential to contaminate usable aquifers; and one-third may endanger water supply wells.

Industrial waste, of course, is not the whole story. Agriculture and mining provide potentially large additional sources of contamination of drinking water.

E.P.A. has recently promulgated cradle-to-grave regulations for hazardous wastes (E.P.A. 1980*b*); however, implementation of these regulations is left primarily to the states, and the magnitude of the source is so large, diverse and complex that compliance will be difficult to accomplish. Equally important, to my knowledge, a quantitative estimate of the magnitude of the current and future risk has not been made, and the recently adopted regulations have not had the benefit of a searching risk–benefit analysis to see where they may require too much or too little.

Estimates exist that the toxicity of hazardous chemical wastes is large compared with that of the radioactive waste generated by a 100 MW_e nuclear reactor (Cohen 1977); I believe that a similar case can be made for its carcinogenicity (Okrent 1980). The difference in public attitude and in regulatory approach to these two kinds of waste is difficult to explain on the basis of quantitative risk.

5. Comparative risk studies

There has been a growing trend towards the use of quantitative risk analysis within many of the federal regulatory agencies in the United States; within the last year or so, pressure for studies of comparative risks and their use in decision-making has arisen within the U.S. Congress. One example of such pressure is a bill (H.R. 4939) introduced by Congressman Don Ritter, himself a metallurgist. The stated purpose of this bill is to provide a federal mechanism for assessing comparative risks involved in actions in scientific, technological and related fields, to apply and promote the understanding and appreciation of comparative risks, and to assist federal, state and local governments, private industry and the public in making intelligent comparisons and evaluations of these risks. The President's Office of Science and Technology Policy (O.S.T.P.) would have been given the authority to carry out the purposes of the act. Although H.R. 4939 was not passed by this session of Congress, hearings on the bill were held on 14–15 May 1980.† The Office of Science and Technology Policy was cool towards the bill, saying that 'proposals to centralize and standardize risk assessment in the federal government are premature, underestimating the long-range success of efforts underway, and overestimating the ability of research and analyses to make decisions'. Representatives of several federal regulatory agencies testified and identified the relatively limited ways in which they are employing risk quantification, and gave little or no indication of the use of comparative risks. None endorsed the bill. Two speakers, representing what is called the 'environmental' point of view in the media, opposed the bill, expressing a fear that the risk evaluation responsibility would put O.S.T.P. into a position 'to influence or perhaps stymie the agencies to whom Congress has given the preventive health mandate'.

Three professors, Howard Raiffa and Richard Wilson of Harvard University, and James Vaupel of Duke University, gave support to the performance of comparative risk assessment. They conceded that such risk assessment was subject to error, uncertainty, or even abuse, but argued that better decisions concerning health and safety priorities are likely to be made if more information is available. I wish to associate myself with their position.

† Congressman Ritter introduced a new bill, H.R. 8303, entitled 'Risk Analysis Research and Demonstration Act of 1980', on 2 October 1980.

In any event, although there is a growing trend towards the use of quantitative risk estimation and assessment, that point of view is by no means unanimous nor is motion always in the same direction. However, I believe that the trend is inexorable.

6. Quantitative safety goals

There have been relatively few published proposals for quantitative risk acceptance criteria for nuclear reactors or for other technological ventures. Many of the existing proposals appear to be subject to rough categorization into one of three groups: those that set limits on individual risk of death only (Adams & Stone 1967; Bowen 1975), those that consider the frequency of accidents and the magnitude of the consequences (Farmer 1967; CIRIA 1977; A.E.C.B. 1978; Kinchin 1978), and those that embed the criteria in risk management frameworks that, at least in part, consider risks from alternative or other societal endeavours (Okrent & Whipple 1977; Comar 1979; Atomic Industrial Forum 1980).

Neither the U.S. Nuclear Regulatory Commissioners (U.S.N.R.C.), nor the U.S. Atomic Energy Commissioners (U.S.A.E.C.) before them, have defined quantitative safety goals or quantitative risk acceptance criteria for nuclear power reactors. In 1965 a kind of quantitative benchmark was provided (implicitly) by the A.E.C. commissioners in connection with their review of the application for a construction permit for a large pressurized water reactor at Malibu, California. The application was contested by a vigorous, well funded intervener group that hired prominent technical consultants who argued that permanent surface displacement might occur on a small fault under the reactor containment building, during a major earthquake on a nearby large fault. The A.E.C. regulatory staff and the applicant for the construction permit, the Los Angeles Department of Water and Power, disagreed. However, the Atomic Safety and Licensing Board, on hearing the case, agreed with the intervener. The A.E.C. regulatory staff appealed the decision to the Atomic Energy Commission. In their ruling upholding the Licensing Board, the Commissioners said that the fact that geological evidence indicated that there had been no displacement along the fault under the proposed containment building in at least 10000 years was not sufficient to rule out a need to design for displacement along the fault. Thus, the ruling provided implicit guidance that reactors should be designed to cope with events having a recurrence interval shorter than 10000 years.

The next suggestion for a quantitative safety criterion for power reactors came in 1973 from the A.E.C. regulatory staff in connection with their report on anticipated transients without scram† (A.E.C. 1973). In this report they defined a safety goal that, assuming a population of 1000 light-water reactors (l.w.rs) in the early 21st century, the frequency of a serious reactor accident should be less than one in 1000 per year from this reactor population (or one in 10^6 per reactor year). The

† Scram failure is a failure of the safety rods to insert rapidly into the core, when needed, to terminate the chain reaction quickly.

regulatory staff defined a serious accident as one that led to doses exceeding those in the siting regulations (A.E.C. 1962), namely 25 rem whole-body or 300 rem to the thyroid.

In 1976, the U.S. Advisory Committee on Reactor Safeguards (A.C.R.S.), in responding to a question from the U.S. Congress, proposed the use, as an interim safety criterion, of 10^{-6} per reactor year for a serious accident, which the A.C.R.S. defined as equivalent to a fatal crash of a loaded commercial aircraft (Okrent 1979).

In May 1979 the A.C.R.S. recommended to the N.R.C. that the N.R.C. should develop quantitative safety goals or risk acceptance criteria for light-water reactors and ask the U.S. Congress to express its views on the suitability of such goals in relation to other relevant aspects of society (A.C.R.S. 1979).

Substantial efforts have been initiated since then; these include various industry groups, a working group of the Institute of Electrical and Electronic Engineers (I.E.E.E.), the N.R.C. regulatory staff and their research contractors, and the A.C.R.S. There is some likelihood that the Congress will require the development of proposed quantitative safety goals by the N.R.C. In any event, the N.R.C. Commissioners have recently approved a plan whose intent is to have the regulatory staff provide specific recommendations for quantitative safety goals to the commissioners before the end of 1981.

7. A trial approach

The A.C.R.S. has had one of its subcommittees examining the matter, and in October 1980 the A.C.R.S. forwarded a trial approach to the N.R.C. to serve as a discussion point in the beginning of an iterative process (A.C.R.S. 1980*a*).

In this proposal (Griesmeyer & Okrent 1980), which draws on past and current work in the field, risk management is divided into two major tasks: the first is the predominantly social and political problem of setting the safety goals, and the second is the technical question of estimating the risks and deciding whether the safety goals have been met. The safety criteria or decision rules are as follows: limits are placed on the frequency of occurrence of certain hazardous conditions (hazard states) within the reactor; limits are placed on the risk to the individual of early death, or delayed death due to cancer, arising from an accident; limits are placed on the overall societal risk of early or delayed death; an 'as low as reasonably achievable' approach is applied with a cost-effectiveness criterion that includes both economic costs and a monetary value of preventing premature death; a small element of risk aversion is applied to infrequent accidents involving large numbers of early deaths compared with a similar number of deaths caused by many accidents, each involving one or two deaths.

Each decision rule on hazard states and on individual and societal risk consist of a pair of numbers: an upper, non-acceptance limit on risk and a lower, safety goal level of risk. Compliance with the upper limit would be required for extended operation of the plant; otherwise, it must be improved within a certain period of time (to be determined) that depends upon the severity of the risk involved. On the other hand, any risk value lower than the safety-goal level would be considered to be in compliance for the particular category of risk. However, risks must be further reduced below these safety-goal levels whenever improvements are possible that meet certain cost-effectiveness criteria for risk reduction. Between the upper non-acceptance limit and the lower safety-goal level of risk is a discretionary range in which case-by-case consideration of uncertainties, regional need for power, and alternative risks is required in the decision as to whether the plant should be allowed to operate for an extended time without modification.

The preliminary numerical values that have been suggested for use in the decision rules are primarily a matter of judgement and are intended to help stimulate discussion and evaluation in concrete terms.

The quantitative values suggested for use in the proposed decision rules are intended to be applicable for new light-water power reactors and may be more stringent than is deemed appropriate for existing plants.

(*a*) *Decision rules*

(i) *Hazard states*

Accidents that damage the facility represent a possible forerunner of more severe accidents. A tentative set of hazard states of progressive severity has been defined and a preliminary set of limits on their rate of occurrence has been proposed, as is shown in table 1. The limit on the frequency of a large offsite release,

Table 1. Limits on occurrence of hazard states

hazard state	probability goal	decision rules on mean frequency: goal level	decision rules on mean frequency: upper limit
significant core damage (> 10 % of noble gas inventory leaking into primary coolant)	less than 1/100 per reactor lifetime	$f_{cd} < 3 \times 10^{-4}$ per reactor year	$f_{cd} < 1 \times 10^{-3}$ per reactor year
large-scale fuel melt (l.s.f.m.) (> 30 % of oxide fuel becoming molten)	less than 1/300 per reactor lifetime	$f_{m} < 1 \times 10^{-4}$ per reactor year	$f_{m} < 5 \times 10^{-4}$ per reactor year
large-scale uncontrolled release from containment given l.s.f.m. (> 10 % of iodine inventory and 90 % of noble gas)	small, given a large-scale fuel melt	$f_{R/m} < 0.01$ per l.s.f.m.	$f_{R/m} < 0.1$ per l.s.f.m.

f_{cd} is the frequency of significant core damage per reactor year.
f_{m} is the frequency of large-scale fuel melt per reactor year.
$f_{R/m}$ is the frequency of large-scale uncontrolled release per large-scale fuel melt.

The upper non-acceptance limits must be satisfied for extended operation of a new plant or for issuance of a construction permit. Between the upper limits and the goal levels is a discretionary range for case-by-case consideration of uncertainties and competing risk. Once the risk level decision rules have been applied, risk must still be reduced if such reduction is reasonably achievable within the cost-effectiveness criterion of table 4.

assuming that a fuel melt has occurred, places emphasis on mitigation as well as prevention of serious accidents. Such a division between accident prevention and accident mitigation is believed to be necessary because of the difficulty in demonstrating with a very high degree of confidence that a frequency of large-scale fuel melts much less than the proposed goal of 10^{-4} per reactor-year can be achieved, in view of the complexities introduced by consideration of matters such as sabotage, earthquakes and other potential multiple failure scenarios.

(ii) *Individual risks*

Equity considerations naturally lead to the notion that an individual should not be unduly burdened by risk. However, the definition of undue risk is complicated by consideration of the risks due to the alternatives, and by controversies over the evaluation of different types of risk.

Individual health and safety risks posed by the light-water reactor include early death and illness, fatal and non-fatal cancers, and genetic effects. It is presumed in this proposal that control of early death and latent cancer deaths will adequately control the other effects as well. For the purposes of the decision rules, the generation of electricity is considered to be an activity important to society, and the risk limits have been set below background risks or those from the principal competing source.

In the United States, girls 10–14 years of age have the smallest death rate, approximately 10^{-4} per year, which is due primarily to accidents. This mortality rate is typical of many occupational risks, and is about two orders of magnitude greater than that posed by risk situations that are normally considered negligible (e.g. lightning). The average death rate for the entire population of the U.S.A. is about 10^{-2} per year.

A 1:2000 chance (0.0005) over an individual's lifetime that the reactors at a particular site will be the cause of the individual's death due to cancer, corresponds to a yearly risk of induction of fatal cancer of about 10^{-5} per year. Since the power generated by a plant is generally beneficial to society, it is suggested here that a 1:2000 or 1:4000 risk of fatal cancer over one's lifetime due to reactors is a plausible goal level. This compares with a background risk of death by cancer due to all causes of 0.15–0.2.

Various methods have been suggested for assigning a weighting factor between delayed cancer death and early death. If the severity of risk is represented by the associated loss in life expectancy, an early death, in which all remaining life expectancy is lost, would, on the average, be 2–3 times worse than a delayed cancer death. On the other hand, some studies in the literature have arrived at a factor as high as 30 for the greater importance of early death compared to delayed death by analysing historical data (Litai 1980) or by noting that death within 1 year seems much worse than death 10–15 years from the present. In the current proposal, a factor of 5 is arbitrarily used between the limit on the risk of delayed cancer death and the limit on the risk of early death.

Table 2 summarizes the proposed decision rules for risk of delayed cancer and early death to the most exposed 'average' individuals.† Note that only a few people would have risks as high as the most exposed individuals, who presumably reside close to the plant site boundaries. Most people would be exposed to risks lower than the goal levels.

TABLE 2. LIMITS ON RISKS TO MOST EXPOSED INDIVIDUAL

	mean frequency per site-year		decision rules on mean frequency per large-scale fuel melt (l.s.f.m.)	
probability goal	goal level	upper limit	goal level	upper limit
probability of delayed death from cancer due to all reactors at a site over lifetime of individual <0.0005	$f_d < 5\times10^{-6}$ per site-year	$f_d < 2.5\times10^{-5}$ per site-year	$f_{d/m} < 0.01$ per l.s.f.m.	$f_{d/m} < 0.05$ per l.s.f.m.
probability of early death due to a reactor accident over lifetime of individual <0.0001	$f_{ed} < 1\times10^{-6}$ per site-year	$f_{ed} < 5\times10^{-6}$ per site-year	$f_{ed/m} < 0.002$ per l.s.f.m.	$f_{ed/m} < 0.01$ per l.s.f.m.

f_d is the individual risk of delayed cancer death per site year.
$f_{d/m}$ is the individual risk of delayed cancer death per large-scale fuel melt.
f_{ed} is the individual risk of early death per site year.
$f_{ed\ m}$ is the individual risk of early death per large-scale fuel melt.
The upper non-acceptance limits must be satisfied for extended operation of a new plant or for issuance of a construction permit. Between the upper limits and the goal levels is a discretionary range for case-by-case consideration of uncertainties and competing risks. Once the risk level decision rules have been applied, risk must still be reduced if such reduction is reasonably achievable within the cost-effectiveness criteria of table 4.

Note that conditional goals are also set for the risk of death, given an accident involving a large-scale core melt. This is again intended to require an emphasis on mitigation of consequences, as well as on prevention of accidents.

(iii) *Societal risk*

It has been suggested in the literature that society is risk-averse when comparing a single, infrequent large accident with a number of small accidents leading to the same total number of fatalities in the same time period. A simple approach, which assesses an equivalent social cost that increases faster than the actual consequences for events involving multiple deaths, uses an equation of the form

$$\text{equivalent social cost} = \sum_{\text{accidents}} (\text{frequency})\,(\text{consequence})^{\alpha},$$

† For the purpose of applying the decision rules, the estimated radiation dose to the most exposed individual would be found by using realistic models, including possible emergency plans. This would be done for all significant accident scenarios. The 'average' individual would be operationally defined as an individual whose response to the dose is the same as the dose response averaged over a representative distribution of the population.

in which α is greater than unity. If α is unity, then the equivalent social cost will be the same as the expected cost (frequency times consequence). Although values of α as high as 2 or 3 have been proposed in the literature for fatalities from accidents, such values would prohibit many existing technological endeavours because of the extremely high equivalent social cost, e.g. dams or large quantities of hazardous chemicals stored close to population centres. We do not believe that society is consciously placing such a high risk-aversion penalty on needed activities, nor that it can afford to (Griesmeyer *et al.* 1979; Okrent 1977).†

TABLE 3. SOCIETAL HEALTH RISK LIMITS

measure of risk	decision rules on societal health risks: goal level	decision rules on societal health risks: upper non-acceptance limit
E_d = the expected value of $\sum_{\text{accidents and normal operation}}$ (frequency) (delayed cancer deaths)	$E_d < 2$ per 10^{10} kWh	$E_d < 10$ per 10^{10} kWh
E_{ed} = the expected value of $\sum_{\text{accidents}}$ (frequency) (early deaths)$^{1.2}$	$E_{ed} < 0.4$ per 10^{10} kWh	$E_{ed} < 2$ per 10^{10} kWh

E_d is the average number of delayed cancer deaths per 10^{10} kWh of electricity generated.
E_{ed} is the average number of equivalent early deaths per 10^{10} kWh of electricity generated.
10^{10} kWh is the amount of electricity generated by a large (1200 MW_e) power plant operating at full capacity for 1 year.

The upper non-acceptance limits must be met for extended operation of a new plant or for issuance of a construction permit. Between the upper limits and the goal levels is a discretionary range for case-by-case consideration of uncertainties and competing risk. Once the risk level decision rules have been applied, risk must still be reduced if such reduction is reasonably achievable within the cost-effectiveness criteria of table 4.

In this proposal it is suggested that the social cost for delayed cancer deaths should be assessed as equal to the expected number of fatalities (i.e. $\alpha = 1$). One estimate of the range of the number of people who die from the pollution arising from a coal-fired plant generating 10^{10} kWh is about 10–200 (Hamilton & Manne 1978), although it may be argued that no proven effects have been unequivocally demonstrated. Ten is proposed here as the upper non-acceptance limit on the delayed cancer deaths due to a nuclear power plant; the goal level is that there be less than two cancer fatalities per 10^{10} kWh.

To provide incentives to reduce the catastrophic potential of accidents, we have tentatively chosen to assess the equivalent social cost of early fatalities by using $\alpha = 1.2$; hence, the equivalent early fatality cost of the plant, E_{ed}, would take the form

$$E_{ed} = \sum_{\text{accidents}} (\text{frequency})\,(\text{number of early fatalities})^{1.2}.$$

† An examination of the relation between large accidents and societal actions suggests that rather than leading to quantitative safety criteria that reflect risk aversion corresponding, say, to $\alpha = 2$, large accidents trigger action by legislative and regulatory bodies that might otherwise have stayed dormant (Baldewicz *et al.* 1974).

The limits on equivalent early deaths are reduced by the same factor of five from the delayed death limits as is used in the limits on individual risk. Table 3 summarizes the decision rules for societal health risks.

(iv) *Societal impact reduction: Alara*

It is proposed to use an 'as low as reasonably achievable' (Alara) cost-effectiveness criterion to judge whether additional risk reduction is required beyond that level of safety required to meet the other decision rules. The cost of an improvement would be balanced against the combined change in the risk of delayed cancer deaths, equivalent early deaths, and economic losses.

TABLE 4. QUANTIFIED ALARA COST-EFFECTIVENESS CRITERIA

expenditure limits for impact reduction

\$1 M per delayed cancer death averted	$\$1\times10^6/(\Delta E_d L)$
\$5 M per early equivalent death averted	$\$5\times10^6/(\Delta E_{ed} L)$
twice the economic loss (due to resource damage) averted:	$2/(\Delta E_r L)$

A particular improvement is 'cost-effective' and required if

$$\text{cost} \leq \{2\Delta E_r + (\$5\times10^6)(\Delta E_{ed}) + (\$1\times10^6)(\Delta E_d)\}\,L.$$

ΔE_d is the change (due to the proposed improvements) in the expected value of

$$\sum_{\substack{\text{accidents}\\ \text{and normal operation}}} (\text{frequency})\,(\text{delayed cancer deaths}).$$

ΔE_{ed} is the change (due to the proposed improvements) in the expected value of

$$\sum_{\text{accidents}} (\text{frequency})\,(\text{early deaths})^{1.2}.$$

ΔE_r is the change (due to the proposed improvements) in the expected value of

$$\sum_{\text{accidents}} (\text{frequency})\,(\text{economic losses}).$$

L is the remaining lifetime of the plant in units of 10^{10} kWh to be generated and the frequencies are calculated per 10^{10} kWh, which is the amount of electricity generated by a large (1200 MW_e) plant operating at full capacity for 1 year.

While there is some limit on the amount that the U.S.A. can afford to spend to reduce risk from all of its technological activities, lest economic instability lead to greater risk directly or indirectly, the current perspective on nuclear reactors may be such that society is willing to spend more for light-water reactor safety than many other things. When cost-effectiveness considerations are employed, the marginal cost limit on expenditure for reducing single fatality risks used by various government agencies ranges from about \$0.1 M per death averted by the U.S. Department of Transportation, \$0.2–2 M per death averted for various analyses by the Consumer Product Safety Commission, up to roughly \$5 M in the N.R.C. application of Alara to routine release of radioactive material (Baram 1980). A much wider range is implied if one looks at regulatory requirements that were implemented without direct consideration of cost-effectiveness (Cohen 1980). It is suggested in this proposal that the marginal cost limit on expenditures be set at $1 M per delayed cancer death averted and $5 M per equivalent early death averted, when 'equivalent' deaths are calculated by using the coefficient $\alpha = 1.2$ for risk aversion.

It is expected that careful study will be required to quantify the economic losses due to property and resource damage. Because of uncertainties and the fact that some impacts cannot be quantified, it is proposed that the marginal cost limit on expenditures to reduce adverse economic impacts should be twice the expected reduction in impact, when applying the Alara criterion. This also stresses prevention rather than repair of possible damage.

Table 4 summarizes the quantified Alara criterion.

(b) *Risk quantification*

The rest of the proposed framework deals with the technical tasks of risk quantification, which will be by no means simple. It has to be acknowledged from the beginning that there will be both large uncertainties in such risk estimates and significant differences between independent estimates of the same risk. The form of the decision rules is intended to compensate in part for some of this uncertainty. Limits are placed on the expected values of the various risks. These expected values are the weighted average of the probabilities; they therefore reflect some of the uncertainties. Also, limits are placed on both the risk of a damaging accident to the fuel and on the risk of a large release of radioactive material, assuming the occurrence of fuel damage, thereby requiring both prevention and mitigation.

A major tool for this effort would be a detailed plant-specific and site-specific quantitative risk analysis that is essentially a probabilistic estimate of the distribution of risks. The details of the analysis would form a safety profile of the particular plant and site that could be used to make risk-based decisions on design or procedural changes. The estimated risk distribution would explicitly express the range of uncertainties and would be used in the application of the decision rules. Special attention must be given to the quality assurance of the risk assessment. There must be a full and explicit identification of the assumptions and limitations of the analysis, and peer review would be required. In addition, it is proposed that a procedure should be established to provide a legally binding determination of those risk distribution values to be used with the decision rules. A possible approach to this aspect is the establishment of a Risk Certification Panel. After peer review of the analyses had been completed, the panel would be given the statutory authority to make a finding on the risk values to be used in the application of the decision rules.

8. A comparison with several technologies

A limited perspective on the preliminary choice of quantitative values for use in the decision rules can be obtained by examining the published estimates of the risks for a few technological situations in terms of the decision rules. For this purpose we have chosen the following: a single 1000 MW_e light-water reactor at an average site in the United States; a hypothetical 1000 MW_e coal-fired plant without special sulphur-removal equipment at a site in the vicinity of Pittsburg, Pennsylvania; the Canvey Island complex; the region around a large coke-oven plant.

Table 5. Range of point estimates for individual risk

technological endeavour	additional annual probability of death to an average individual in the most exposed group (non-occupational)
hypothetical light-water reactor at average site in U.S.A.	3×10^{-7} to 3×10^{-5}
hypothetical coal plant	1×10^{-5} to 2×10^{-4}
Canvey Island	3×10^{-5} to 1×10^{-3}
U.S. coke-oven facilities (the 100000 most exposed individuals)	3×10^{-5} to 10^{-4}
proposed goal level (upper limit) for early death	1×10^{-6} (5×10^{-6}) per site-year
proposed goal level (upper limit) for delayed death	5×10^{-6} (2.5×10^{-5}) per site-year

Table 6. Range of point estimates for societal risk (non-occupational)

technological endeavour	risk per year (deaths)
hypothetical light-water reactor at average site in U.S.A.	0.02–2.0 (early plus latent cancer deaths)
hypothetical coal plant	10–200
Canvey Island	7–11
coke-oven facilities in U.S.A.	*ca.* 150
proposed goal level (upper limit) for a light-water reactor	2 (10)

For the light-water reactor, an estimated range of health effects was obtained by using the median estimates of a slightly modified version of the Reactor Safety Study, WASH–1400 (N.R.C. 1975), or an assumed situation wherein the accidents are 100 times more probable.

For the coal plant, the societal health effects are based on a study by Morgan *et al.* (1978*a*, *b*). The individual risk numbers range from the average estimated risk to those living within an 80 km radius of a single plant to the integrated effect of large numbers of coal plants. Coal appears to be different from nuclear or chemical installations in that the individuals most at risk from coal plants may live far from any individual plant and will have a risk exposure roughly proportional to the total number of such plants upwind of them.

The Canvey Island numbers are those estimated to residents from the industrial complex before making the suggested safety improvements.

The coke-oven estimates are taken from a preliminary report of the Carcinogen Assessment Group of E.P.A. (1978).

The individual risk numbers are summarized in table 5, while the societal risk numbers are summarized in table 6 (Johnson & Kastenberg 1980).

Table 7. Effect of risk aversion

(Factor by which expected number of deaths would be multiplied according to power law model of risk aversion for Canvey Island.)

risk aversion factor, α . . .	1	1.2	1.5	2
multiplicative effect	1	5–6	65–75	4500–6000

A brief indication of the difference in equivalent social cost which results from various possible choices of a risk aversion exponent, α, in the power law model is illustrated in table 7 for Canvey. Clearly, at $\alpha = 2$ the calculated equivalent social cost is huge compared with the expected value. For $\alpha = 3$, an accident taking 1000 lives would have an equivalent social cost of 10^9 lives. It seems highly unlikely that society expects or demands that the probability of severe accidents be reduced in accordance with a risk aversion effect equivalent to a value of $\alpha > 1.5$, if that much. Even with a smaller risk aversion, corresponding to $\alpha = 1.2$, it seems unlikely that the simple power law has general applicability, either for nuclear or non-nuclear technologies. Particularly for very improbable events posing very large consequences, such as the potential for $1\text{–}2 \times 10^5$ deaths due to gross failure of a dam, a societal decision must be made, on other bases, concerning the essentiality of the venture, and then an approach to optimization of the safety design must be developed.

It is of some interest to note that large numbers of people are estimated to suffer an incremental risk of premature death equal to or greater than 10^{-4} per year of exposure from technologies such as that involving coke ovens or the burning of coal to generate electricity. Clearly, segments of the public are exposed to still higher risk from some ill-controlled local sources of air and ground pollution, e.g. specific lead smelters, asbestos mines or chemical dumps.

There appears to be a need for a much more general effort towards the quantitative estimation of industrial risk, as practical, so that such information can serve as an input into societal decision making.

References (Okrent)

A.C.R.S. 1979 Letter to Hon. Joseph M. Hendrie from Max W. Carbon, on Report on Quantitative Safety Goals, 16 May.

A.C.R.S. 1980*a* *U.S. Nuclear Regulatory Commission report*, no. NUREG–0739.

A.C.R.S. 1980*b* Letter to Hon. Victor Gilinsky from Milton S. Plesset, on Technical Basis for Risk Comparison with Other Methods of Electricity Generation, 7 May.

Adams, C. A. & Stone, C. N. 1967 In *Proceedings for a symposium on the containment and siting of nuclear power plants*, pp. 129–142. Vienna: International Atomic Energy Agency.

A.E.C. 1962 *Fed. Register* **27**, 3509–3511.

A.E.C. 1973 *U.S. Atomic Energy Commission report*, no. WASH–1270.

A.E.C.B. 1978 *Atomic Energy Control Board report*, no. AECB–1149.

Atomic Industrial Forum 1980 *Statement on the use of probabilistic risk assessment in the regulatory process*. Washington, D.C.: Atomic Industrial Forum Committee on Reactor Licensing and Safety.

Ayyaswamy, P., Hauss, B., Hsieh, T., Moscati, A., Hicks, T. E. & Okrent, D. 1974 *University of California, Los Angeles, Publ.* no. UCLA-ENG-7423.

Baldewicz, W., Haddock, G., Lee, Y., Prajoto, Whitley, R. & Denny, V. 1974 *University of California, Los Angeles, Publ.* no. UCLA-ENG-7485.

Baram, M. S. 1980 *Ecol. Law Qt.* **8**, 473–531.

Bowen, J. 1975 In *University of California, Los Angeles, Publ.* no. UCLA-ENG-7598, pp. 581–590.

CIRIA 1977 *Rationalization of Safety and Serviceability Factors in Structural Codes*, Rept no. 63. London: Construction Industry Research and Information Association.

Cohen, B. L. 1977 *Rev. mod. Phys.* **49**, 1–20.

Cohen, B. L. 1980 *Hlth Phys.* **38**, 33–51.

Comar, C. L. 1979 *Science, N.Y.* **203**, 319.

Comptroller General 1979 *U.S. General Accounting Office Publ.* no. CED 79–30.

E.P.A. 1978 *Carcinogen Assessment Groups' preliminary report on population risk to ambient coke oven exposure.*

E.P.A. 1980*a* *Planning workshops to develop recommendations for a ground water protection strategy*. Washington, D.C.: U.S. Environmental Protection Agency.

E.P.A. 1980*b* *Fed. Register* **45**, 33066–33588.

Farmer, F. R. 1967 In *Proceedings of a symposium on the containment and siting of nuclear power plants*, pp. 303–329. Vienna: International Atomic Energy Agency.

Griesmeyer, J. M. & Okrent, D. 1980 In *U.S. Nuclear Regulatory Commission report* no. NUREG–0739.

Griesmeyer, J. M., Simpson, M. & Okrent, D. 1979 *University of California, Los Angeles, Publ.* no. UCLA-ENG-7970.

Hamilton, L. & Manne, A. S. 1978 *I.A.E.A. Bull.* **20**, 44–58.

Health and Safety Executive 1978 *Canvey: an investigation of potential hazards from operations in the Canvey Island/Thurrock area*. London: H.M.S.O.

Johnson, D. & Kastenberg, W. E. 1980 In *U.S. Nuclear Regulatory Commission report*, no. NUREG–0739, pp. 77–151.

Kinchin, G. H. 1978 *Proc. Instn civ. Engrs* **64**, 431–438.

Litai, D. 1980 Ph.D. thesis, Massachusetts Institute of Technology.

Morgan, M. G., Rish, W. R., Morris, S. C. & Meier, A. K. 1978*a* *J. Air Pollut. Control Ass.* **28**, 993–997.

Morgan, M. G., Morris, S. C., Meier, A. K. & Shenk, D. L. 1978*b* *Energy Syst. Policy* **2**, 287–310.

N.R.C. 1975 *U.S. Nuclear Regulatory Commission report*, no. WASH–1400 (NUREG–75/014).

Okrent, D. 1977 *University of California, Los Angeles, Publ.* no. UCLA-ENG-7777.

Okrent, D. 1979 *On the history of the evolution of light-water reactor safety in the United States.* Los Angeles: University of California, Los Angeles School of Engineering and Applied Science.

Okrent, D. 1980 In *Engng Foundation Workshop on Risk/Benefit Analysis in Water Resources Planning and Management*, 21–26 September. (In the press.)

Okrent, D. & Whipple, C. 1977 *University of California, Los Angeles, Publ.* no. UCLA-ENG-7746.

Seed, H. B. & Lee, K. L. 1973 In *San Fernando, California, earthquake of February* 9, 1971 (ed. N. A. Benfer & J. L. Coffman), vol. 1, pt B, pp. 809–813. Washington, D.C.: National Oceanic and Atmospheric Administration.

Slesin, L. & Ferreira, J., Jr 1976 *Social values and public safety: implied preferences between accident frequency and severity*. Cambridge: Massachusetts Institute of Technology Laboratory of Architecture and Planning.

U.S. Bureau of Reclamation 1978 *Summary report: water level restrictions at Jackson Lake, Wyoming* – 1978 *decision basis*.

Discussion

H. J. Dunster (*Health and Safety Executive, London, U.K.*). I was surprised to see how close the 'unacceptable' and 'target' figures are. In general, they differ only by a factor of 5. In my book that makes them identical.

D. Okrent. I agree that a factor of 5 difference between the 'unacceptable' and 'target' figures is within the range of uncertainty. The real question posed by this proposal in this regard is, 'should there be two levels of risk?' The numbers given are intended to suggest the range in which such numbers might lie. If the numbers are large enough, a factor of 5–10 in the average value may be usable, despite the large inherent uncertainties. However, this remains to be evaluated.

G. Van Reijen (*Commission of the European Communities, Brussels, Belgium*). I appreciate very much the way in which Professor Okrent proposes to break the isolation of nuclear energy in the context of safety by comparison with non-nuclear activities. In the U.K., the disadvantages of the isolation mentioned have been realized earlier than in most other countries, when in 1975 the Nuclear Installations Inspectorate became an integral part of the Health and Safety Executive. Doubts can be expressed on the proposal, mentioned by Professor Okrent, to establish goal levels and upper limits for certain risks. In particular in the nuclear field, the following goal levels for the most exposed individuals have been indicated in his lecture: for delayed death, 5×10^{-6} per site year; for early death, 1×10^{-6} per site year. I see some discrepancy between the difference of only a factor of 5 between both figures given here and the fact that in the Reactor Safety Study (N.R.C. 1975) the risk of delayed death is evaluated to be a factor of 300 higher than the risk of early death.

D. Okrent. The relative numerical values proposed on a trial basis as safety goal levels of risk to the individual were not established on the basis of the predictions of studies like the Reactor Safety Study. Rather, a plausible number was sought for the safety goal for the risk of early death. Then, partly in comparison and partly from other considerations, a plausible safety goal was sought for the risk of delayed death (due to cancer induced by radiation from an accident, in this case). One of these goals may prove to be limiting for a nuclear power reactor, while the other might be limiting for a different technological situation, were the same safety goals to be applied to it.

Proc. R. Soc. Lond. A **376**, 151–165 (1981)
Printed in Great Britain

Risks in the built environment

By M. W. Holdgate
Departments of the Environment and Transport,
2 Marsham Street, London SW1P 3EB, U.K.

There are good statistics for deaths in transport accidents, fires, and from accidents in the home in Great Britain, and considerable (but less comprehensive) information about injuries and material damage. Information about the causes of these events is much more scanty, and little is known about the long-term effects of accidental injury. The available data are reviewed and the nature, magnitude and frequency of various kinds of risks are analysed for different age groups and in relation to environmental and other factors. The contribution of 'voluntary' actions (notably alcohol and smoking) is assessed. Finally, the extent to which both actual and perceived risk can be modified by education, engineering (modifying the design of roads, vehicles, aircraft, homes and fittings), and the enforcement of regulations and control systems is evaluated. While costs cannot easily be measured, false perceptions of risk can lead to wasteful investment, and education and information are essential if resources are to be deployed where they will do most good.

Introduction

Content and definitions

This paper discusses some of the risks encountered by individuals in everyday living and travel in an urbanized industrial society. It is not concerned with the risks to which populations may be exposed as a result of major civil engineering failure (e.g. of dams or sea walls). Nor is it concerned with risks at work.

Risk is here defined as the probability of a particular outcome following from an event or action (where this cannot be quantified, the situation is one of *uncertainty*). In relation to road traffic, for example, it may be stated as the number of accident occurrences related to a measure of exposure, the latter having both spatial and temporal dimensions. *Risk estimation* is the process of identification and measurement of these probabilities: *risk evaluation* is a broader process of judgement of the social significance and acceptability of such risks. Risk evaluation naturally demands some examination of the factors underlying the risk: again, referring to road accidents, the approach is to identify those without which the accident would have been less likely to happen, or to have been less serious, to assess their relative importance and to define their interactions.

The adequacy of the data

Any estimation of risk thus clearly depends critically upon the type and quality of available data. There is a considerable volume of information about risks in transport. The Department of Transport publishes an annual digest of statistics: these record trends in road, rail and air accidents leading to deaths or injuries (Department of Transport 1978). Rail accidents are recorded in the reports of the Chief Inspector of Railways (see, for example, McNaughton 1980).

National statistics for risks in road transport are based on police reports and cover all fatal accidents and approximately 70% of injury accidents. In England, Wales and Scotland these data are collated by the Department of Transport: a similar system operates in Northern Ireland. The statistics record the location, time and circumstances of each accident and the types of road user and vehicle involved, and allow trends to be established. These records are extended by a National Hospital In-Patients Enquiry collated by the Office of Population Census and Surveys (O.P.C.S.), giving clinical information about a 10% sample of road traffic accident casualties detained in hospital, and by Coroners and Procurators Fiscal fatality reports on adult deaths, sent to the Transport and Road Research Laboratory (T.R.R.L.), which cover about 45% of all fatalities and are particularly informative about blood alcohol levels. In addition, regional information is provided through local accident investigation units (police and highway engineers), which give greater detail about accident situations and factors, and these are extended by in-depth investigations done or supported by T.R.R.L. (T.R.R.L. 1978).

Fire risk is another area of great social concern. The Home Office publishes each year statistics for all fires attended by U.K. fire brigades (but not by factory or other private teams) (Home Office 1978). There can only be crude estimates (e.g. from insurance claims) of the number of small fires not attended by brigades, but some research suggests that they outnumber the others by a factor of up to 5 (Crossman *et al.* 1977). Details of types of injury (due to gas, smoke, burns, scalds or shock) are recorded in the official statistics, as are types of building and occupancy, time and place, source of ignition, material ignited, and extent. The fatality data collected by the Registrar General agree broadly with fire brigade records.

Financial data are, however, available only as estimates of monthly losses from the British Insurance Association, which suggest that in Britain total fire losses amount to 0.2–0.25% of the Gross Domestic Product. Statistical data are available for all accidental deaths in homes and residential institutions in England and Wales (O.P.C.S. 1978). The primary causes of death (e.g. fall, fire, poisoning, electrocution) are recorded, as are the age and sex of the victims and the month of occurrence. There are no data on how far these deaths are associated with structural failure in buildings, though other data imply that this is a minor factor for most fire deaths. The information is extended by the Home Accidents Surveillance System (H.A.S.S.) which has, since 1977, collected data from 20 hospitals on accidents in the home leading to treatment. Again, types of injury, age, sex, and the articles or features in the home associated with the incident are recorded. The Building Research Establishment (B.R.E.) has reclassified the information so as to explore its relation to building features such as stairs or architectural use of glass. Data are also available on injuries or deaths due to gas explosions, damage to buildings from colliding vehicles, wind damage and structural collapse.

Reviewing the data as a whole, it is probably true to say that statistics for fatal accidents are comprehensive, but those for injury imperfect. There are three reasons for this. First, society is understandably concerned to record and investigate events leading to death. Secondly, the total number of injuries is so great that it would be impracticable to record details of a 100% sample. Thirdly, many injuries do not lead to hospital treatment and may be unnotified and unrecorded. The consistency of the records has also been affected by changes in the recording system used: differences in recording practice also make valid comparisons between U.K. fire casualty statistics and those in other countries very difficult (Home Office 1980). Data on causation of accidents are much less adequate (which is not surprising since only thorough investigation at the scene can provide a reliable judgement): in most areas of risk the best that can be done with the statistics is to establish correlations between accident rates and certain recorded factors (like age, season or weather conditions). A further problem arises where many factors contribute to an accident but only the immediate cause of death is recorded (fatalities among the elderly through falls are clearly the consequence of frailty as well as staircase design). The data also generally fail to cover deferred effects such as chronic ailments leading to a shortening of life, although some research has been done on permanent disability in road traffic accident casualties (Grattan & Hobbs 1980), and long-term effects on the health of firemen are being studied (Home Office 1979). Information about economic losses due to material damage is the most defective of all, making any calculation of social cost, important in risk evaluation, extremely uncertain.

The nature, magnitude, frequency and perception of risk

Risks in transport

Table 1 records the fatal and non-fatal injuries in road, rail and air travel in the United Kingdom from 1968–78. It is clear that road accidents account for the vast majority. Table 2 breaks the data for the latter in 1977 down by class of road user and road type. The familiar facts that motor cyclists run more than twice the risk of fatal injury of pedal cyclists, and twenty times the risk of car drivers, that commercial vehicles and public service vehicles are safer still to travel in, and that motorways are the safest roads to travel on, are all immediately apparent. Occupational risks are clearly low compared with public risks.

Most road accidents involve only one or two vehicles. Multiple pile-ups on motorways are rare. In 1977 only 46 deaths occurred in 33 accidents involving 3 or more vehicles in Great Britain. In that year there were less than 17 cases of coach crashes, leading to 34 fatalities. Yet, as in other circumstances, these catastrophic events attracted more public concern (if press publicity is to be taken as a measure of this) than the deaths of 6500 other road users in the same year (Sabey & Taylor 1980). Rather similarly, as table 1 confirms, air travel (measured in terms either of gross fatalities per 10^5 travelling or of fatalities per unit of distance travelled) is about six times safer than road travel (Stratton 1974). Yet because a high proportion of aircraft accidents have no survivors (57 % of 1713 accidents to public transport aircraft in 1946–73) and because of the number of fatalities incurred in many individual incidents (37 accidents with over 100 victims in 1970–8), the public perception of flying, like that of motorway driving, as a dangerous form of travel has been sustained.

Table 1. Road, rail and air casualties in Great Britain, 1968–78

(Department of Transport (1978).)

	1968	1969	1970	1971	1972	1973	1974	1975	1976	1977	1978
road											
deaths	6810	7365	7499	7699	7763	7406	6876	6366	6570	6614	683
injured	342398	345529	355869	344328	351964	346374	317726	318584	333103	341447	34296
all road casualties	349208	352894	363368	352027	359727	353780	324602	324950	339673	348061	34979
rail											
deaths†‡§	334	352	381	373	328	306	360	416	382	401	44
injured†‡§	6497	7114	6692	6390	6245	6464	6218	5725	5304	5301	575
all rail casualties	6831	7466	7073	6763	6573	6770	6578	6141	5686	5702	620
air flights‖											
deaths‖	53	—	—	63	118	—	—	—	63	—	—
injured¶	40	8	5	4	20	2	4	9	—	—	
all domestic air flight casualties	93	8	5	67	138	2	4	9	63	—	

† Includes staff employed by National Carriers and Freightliners Ltd.
‡ Includes all killed and injured except railway staff on railway premises.
§ Includes suicides and attempted suicides.
‖ On scheduled passenger services of U.K. airlines.
¶ Requiring hospital treatment.

Rail is also a low-risk form of travel (table 1). The accident rate has improved markedly over time. For example in 1920, 1930 and 1970 there were 0.6, 0.4 and 0.1 collisions respectively per 10^6 train miles (excluding impacts on buffer stops). Total fatalities in these years from movement accidents were 80, 50 and 30 respectively. Yet public reactions to rail accidents, as to air accidents, is strong, perhaps for two reasons. First, some accidents have involved large numbers of casualties and hence, like aircraft crashes and motorway pile-ups, have a special psychological impact. Secondly, rail and air transport are public services, and are expected to be safer than the transport that the individual provides for himself.

Most transport accidents have a multiplicity of causes (or of contributory factors), and this is well illustrated by reference to the roads (Sabey & Taylor

1980). Figure 1 illustrates, however, that neither the road environment nor the vehicle is usually the sole cause of an accident, and that overwhelming responsibility rests with the road user. Perceptual errors (imperfect observation, inattention or misjudgement) and errors in the execution of a manoeuvre account for 60 % of these incidents. Bad road layout, poor maintenance, or inadequate markings account for a high proportion of the contributory environmental factors. In addition, risk is considerably modified by individual 'voluntary' behaviour. The

TABLE 2. ROAD CASUALTY RATES IN GREAT BRITAIN IN 1977

(Sabey & Taylor (1980).)

(*a*) *for different road users*

user	casualties per 10^8 occupant kilometres		
	killed	serious injury	all injuries
motorcyclists	15.9	274	966
pedal cyclists	6.8	107	531
car occupants	0.7	9	42
commercial vehicle drivers			
less than $1\frac{1}{2}$ t u.w.	0.5	7	31
more than $1\frac{1}{2}$ t u.w.	0.4	4	16
public service vehicles	0.1	2	21
all motor vehicle users	0.8	12	52

(*b*) *for different road types*

user	casualties per 10^8 vehicle kilometres	
	fatal	all injuries
motorways	0.84	26
rural roads (50–70 mile/h† speed limit)		
'A' class	2.9	79
others	1.9	89
urban roads (30–40 mile/h† speed limit)		
'A' class	3.1	191
others	2.5	192

† 1 mile/h ≈ 1.6 km/h.

consumption of alcohol is the most serious such factor. Since 1966, blood alcohol levels have been recorded by coroners in England and Wales (and in Scotland since 1978 by the Procurators Fiscal) for road fatalities aged 16 or over. Table 3*a* indicates that 32 % of all drivers of three-wheeled and four-wheeled vehicles, and 29 % of all motorcycle drivers, killed in 1978 had blood alcohol levels over the legal limit. Table 3*b* demonstrates that on Friday and Saturday nights the percentage rose to over 60 %. The table also shows that a significant proportion (well over 10 %) of all road users killed had blood alcohol levels exceeding 150 mg per

100 ml, and figure 2 demonstrates how the proportion above the legal limit has risen in recent years. Like the failure to attach a seat belt, or smoking in bed, identified as a significant cause of deaths in fire, this can be described as a voluntarily undertaken risk but it is one that also imposes risks on others and major costs on the community.

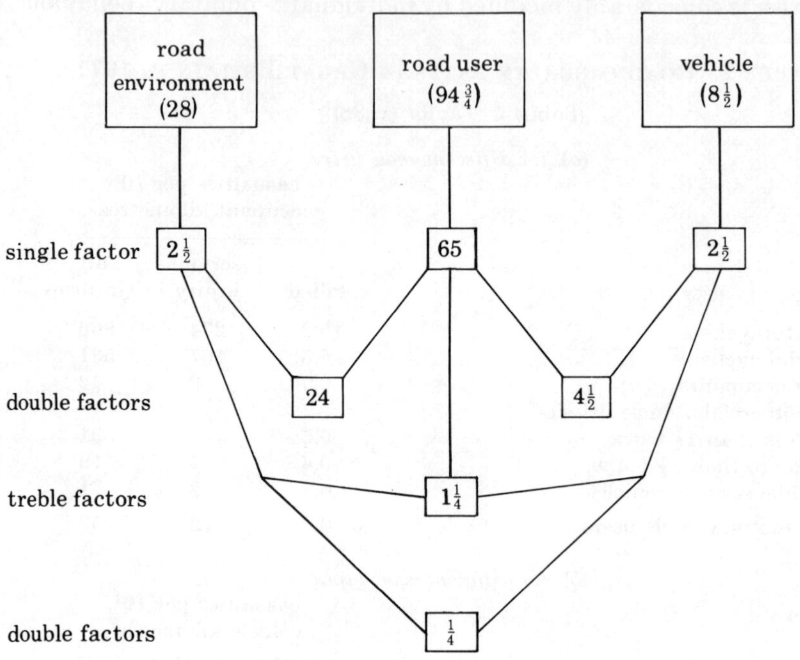

FIGURE 1. Factors contributing to road accidents (percentage contributions; the total percentage contribution for each factor is shown in parentheses) (Sabey & Taylor 1980).

There is evidence that the perception of risk by drivers is far from reliable, and directly contributes to accidents on the road (Watts & Quimby 1980). Table 4 lists some commonly overrated and underrated hazards. More generally, it has been pointed out that for most people, travel, even by road and by the more hazardous modes, is perceived as relatively safe. The risk of involvement in an accident involving injury on the road in Great Britain is once in 57 years, of a fatal accident 1 in 2500 years, and of an accident not involving injury about 1 in 9 years. Even though one random sample of drivers expressed the belief that they were likely to be involved in a serious accident in the future (Sheppard 1975), rates of this kind could well explain why individuals tend not to regard road safety as a matter of pressing importance for them. It may also explain why, although it is a widely publicized fact that wearing seat belts approximately halves the risk of serious injury to occupants in a car accident, many people do not follow this practice – perhaps because of a subjective fear of being trapped in

ION OF BLOOD ALCOHOL CONCENTRATION (b.a.c.)
USERS KILLED IN GREAT BRITAIN IN 1978
abey & Staughton (1980).)

(*a*) *by b.a.c. levels*

percentage with b.a.c. (mg/100 ml) exceeding:							
9	50	80	100	150	200	number in sample	all fatalities aged 16 or over†
44	37	33	31	20	11	647	1732
42	35	29	25	14	6	396	1002
49	37	32	29	14	8	352	1230
34	28	25	23	19	13	571	1940
21	19	16	16	16	7	58	202
41	**34**	**29**	**27**	**17**	**10**	**2024**	**6106**

† Approximately 75% of these died within 12 h of the accident.

(*b*) *by time of day*

percentage with b.a.c. exceeding 80 mg/100 ml between 10 p.m. and 4 a.m.

	Friday/Saturday and Saturday/Sunday nights	all nights
drivers	68	64
riders	62	55

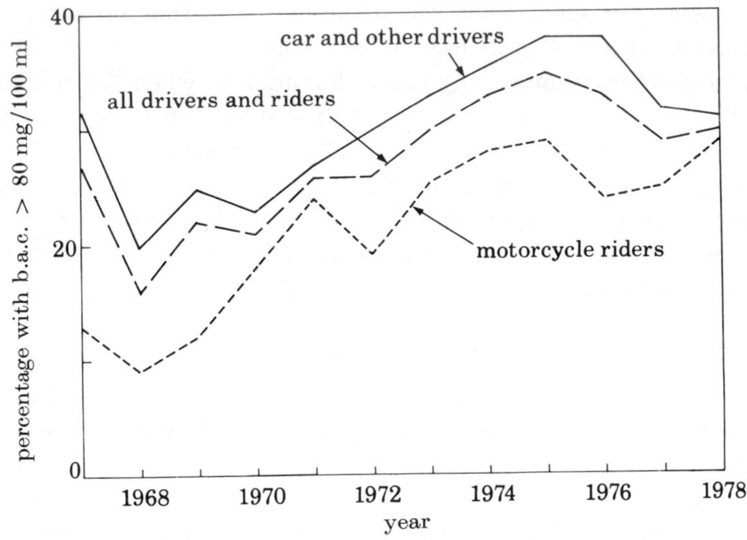

FIGURE 2. Trends in blood alcohol concentration (b.a.c.) in fatalities in England and Wales (Sabey & Staughton 1980).

a burning or overturned car, something that happens in only a tiny proportion of accidents. Actual experience is simply insufficient for individuals to perceive the true balance of risk for themselves.

Table 4. Ranking of underrated and overrated hazards in road travel

(Sabey & Taylor (1980).)

description of hazard	difference in rank
underrated	
suburban dual carriageway near a pedestrian bridge	−26.0
a rural crest on a single carriageway	−25.5
a left turn off a rural road	−25.0
a rural dual carriageway near a picnic area	−24.5
rural cross-roads controlled by traffic lights	−21.0
overrated	
hump bridge on rural road	+35.0
level crossing on rural road	+32.0
suburban shopping centre	+31.0
right turn onto rural dual carriageway	+24.5
right bend at the end of rural dual carriageway	+22.0

Risks in the home

About 7000 people die each year in Britain as a result of accidents in the home. Calculations based on H.A.S.S. data suggest that about 10^6 accidents lead to hospital treatment. Two-thirds of these accidents are not directly related to the features of the homes or buildings. The pattern of accidents is unevenly spread with age, and especially affects the elderly. Very few of these deaths and injuries are associated with the spectacular catastrophic events that attract public concern. Only 13 structural collapses have been recorded in the past 20 years, most of them in places where people congregate, fortunately when the buildings were unoccupied. Only four fatalities resulted from these incidents, all at Ronan Point (W. H. Ransom, private communication). Risk of deaths or injury from wind damage – falling chimney stacks, overturned caravans, etc. – is also small, totalling about five cases a year. Gas explosions and vehicle collisions with buildings (taken together) are somewhat more frequent, averaging one per day, one significant incident a week, and one fatality per month.

Risks from fire

Table 5 shows the location of fires attended by brigades in 1977. In 1976, fires caused 1.6 deaths and 11 injuries per 10^5 population. For comparison, all accidents in the home (including fires) accounted for 11 deaths per 10^5 and about 1600 injuries and road accidents 13 deaths per 10^5 and 613 injuries. The risk of death in fire is about equal to that due to accidental poisoning, industrial accident, homicide and assault or accidental drowning (Home Office 1980). Casualties show a marked variation with age with the greatest risk of fatality in the very old and the very young (although the latter have a below-average risk of injury) (table 6). The fact that the under-5 age group has twice the casualty rate of the 5–14 years group may reflect the fact that school is safer than home (no children had died in a school fire during school hours in the U.K. since modern records began).

TABLE 5. THE LOCATION OF FIRES IN 1977

(Home Office (1980).)

dwellings	51 000
industrial buildings	11 000
shops	3 500
public entertainment, clubs, pubs, etc.	3 800
agricultural premises	3 000
hospitals	3 100
schools	1 800
hotels, etc.	1 700
unknown and unrecorded	15 500
derelict buildings	16 000
confined to chimneys	46 000
grass land, refuse unrecorded	139 400
road vehicles	24 300
other outdoor fires, ships, caravans, etc.	7 400

TABLE 6. FIRE CASUALTIES BY AGE GROUP, U.K., 1976

(Home Office (1980).)

age/year	fire casualties per 10^6 persons	fire deaths per 10^6 persons	percentage of casualties resulting in death
under 5	100.0	21.1	21.1
5–14	53.1	7.1	13.5
15–59	138.0	10.1	7.3
60–74	109.9	23.0	20.9
over 74	230.7	81.8	35.5
all ages	122.2	15.9	13.0

As with deaths in transport and buildings, public concern is greatest when fires cause multiple deaths, especially when these tragedies involve children or the helpless aged. In 1977, out of 609 reported fatal fires, 54 caused more than one death, with an average of 3.5 fatalities per incident; 47 of these fires occurred in dwellings. One-third involved the deaths of children under 15 years alone, and two thirds, only children under 15 years and adults under 30 years. A high proportion occurred at night. The multiple fatality fires with the greatest number of victims in recent years have, however, occurred in old people's homes and nursing homes, with up to 18 victims in one incident. Fires attracting special public concern therefore appear to form a higher proportion of the total number of incidents than do multiple-casualty accidents on the roads or in the home.

Figure 3 indicates the range of causes of all fires. Cooking and space heating are the main causes of non-fatal fires. Smoking in bed makes a disproportionately large contribution to the fatality figures in both the United Kingdom and the

United States. As with the accentuated risk from drinking, which is known to be a contributory factor in fire risk to adults (Harland & Woolley 1979), this is to some extent a hazard voluntarily incurred – but it involuntarily hazards others. Other changes in hazard in recent years are related to trends in materials and furnishing fabrics, with the increasing use of plastic foams, which are more flammable than traditional materials and also create asphyxiating or toxic fumes

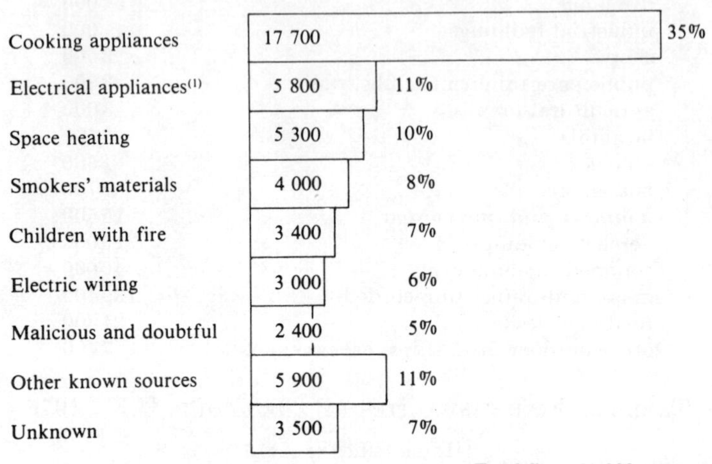

FIGURE 3. Fires in dwellings by source of ignition, U.K., 1977 (Home Office 1980).

when smouldering. The more rapid spread of fire from these materials, coupled with decreasing mobility among ageing population groups, considerably increases the risk of fire fatality. This may be one reason why the proportion of fatalities attributable to asphyxiation by gas or smoke has risen over the last two decades, and now appears to have levelled off at around 50 %. Regulations and advice to mitigate the risks involved are discussed in the following section.

Risk management

In all the fields reviewed in this paper, efforts have been made to reduce risk by management and information. In transport, counter-measures can be grouped in three categories: engineering (of road, track, signals and communications or vehicles), education, and enforcement. Because the risks in rail and air travel are already relatively low, the greatest potential for action lies on the roads (although all reasonable steps toward gains in the other two modes will clearly be taken). It has been calculated that road accidents in Britain cost some £1600M annually (although the assumptions about the cost of pain, grief and suffering and the evaluation put on life are clearly arbitrary and controversial) (Sabey & Taylor

1980). The costs of road safety measures are put at £1000 M. Nonetheless, there are clear economic incentives as well as social ones for further action.

Sabey & Taylor (1980) suggest that better road engineering might prevent 20 % of accidents, and that some of these measures are so low in cost that the economic rate of return in the first year could be as high as 400 % (although 50 % is a more reasonable average) (see table 7). In round terms, expenditure of £100M spread

TABLE 7. POTENTIAL FOR ACCIDENT AND INJURY REDUCTION IN ROAD ACCIDENTS, BASED ON 1977 DATA

(Sabey & Taylor (1980).)

options	potential: percentage savings
road environment (*low-cost remedies*)	
geometrical design, especially junction design and control	10½ (11½)
road surfaces in relation to inclement weather and poor visibility	5½
road lighting	3 (1½)
changes in land use, road design, and traffic management in urban areas	5–10 (7½–16½)
overall	one-fifth of accidents
vehicle safety measures	
primary	
vehicle maintenance, especially tyres and brakes	2
anti-lock brakes and safety tyres	7 (6)
conspicuity of motorcycles	3½ (3)
secondary	
seat belt wearing	7 (10)
other vehicle occupant protection measures	5–10
overall	one-quarter of casualties
road user and road usage	
restrictions on drinking and driving	10
more appropriate use of speed limits	5
propaganda and information	up to 5
enforcement and police presence	up to 5
education and training	up to 5
other legislation (e.g. restrictions on parking)	up to 5
overall	one-third of accidents

Figures in parentheses indicate earlier values based on 1973 data, where these are different from latest estimates.

over a number of years in a comprehensive low-cost road engineering programme (for example improving junction design and control, road surface texture and road lighting) might bring a return of £250M in accident savings. Vehicle engineering measures (better maintenance, anti-lock brakes, safety tyres and making motorcycles more conspicuous), together with more universal seat belt wearing and other measures to protect occupants, might prevent 25 % of casualties overall. In Australia, the enforcement of seat belt regulations is said to have cut deaths and injuries by 20–40 % overall for car occupants (Grime 1979), and in the U.S.A.

the use of air bags and passive seat belts has produced between 50 and 70% reductions in death rates (Department of Transportation 1978).

Education and enforcement together, through restrictions on drinking and driving, appropriate speed limits, education and training of children and young adults as pedestrians, pedal cyclists and motorcyclists, publicity and information campaigns and various other measures like parking restrictions, are assessed as having the potential to reduce accidents by as much as 30% – and one calculation suggests that a combination of legislation and publicity to curb alcohol consumption among drivers, if it brought a 10% saving in accidents, would save £150M annually at current prices and incidentally save 100000 hospital bed-nights annually. If the enforcement of legislation and publicity together cost £10M annually, the economic returns thus emerge as of the order of 10:1 (Sabey 1980).

'Engineering' methods have also been employed to effect reductions in fire risk, for example through changes in the design of shopping malls to reduce the risk of rapid spread of smoke, and through better layout of other public buildings and installation of systems to control air and smoke movement. On a smaller scale, changes in the design of paraffin heaters led to detectable falls in fire incidence in the home. Better ways of glueing polystyrene tiles to a ceiling so that they burn *in situ* rather than fall, spreading flame to the floor, have been devised and publicized. Better detection and warning systems have an increasing part to play in the future. Building regulations can assist in ensuring that design and construction methods avoid the needlessly hazardous, as can regulations on flammability for furnishings (recently introduced) and measures under the Fire Precautions Act and other older Acts providing for certification and inspection of certain kinds of premises. Short-duration educational measures (like 'fire weeks') have had short-term effects (barely detectable statistically) but not longer-term ones. There is likewise no evidence that campaigns to improve safety in the home have had any effect on accident rates. Some such accidents could be reduced by engineering means (e.g. the use of safety glass instead of annealed glass), but the extra cost would not easy be easy to justify when it is remembered that about 66% of accidents in the home do not involve a building feature.

Conclusions

The built environment, in the sense used in this paper, is everyone's habitat in our industrialized society. It is not surprising that various risks attach to it. This paper has demonstrated the relative magnitude of risk in the different transport modes and in the home. Fire risks have been given particular attention. Taken together, these hazards account for some 25 deaths per 10^5 of the population in the U.K. per year. The risk of dying from one or other of them is therefore about one-tenth of that of dying from cancer or one-quarter of that of dying of pneumonia. The risk of injury is some two orders of magnitude greater. But these general statistics obscure a considerable variability in risk between sectors of the population, and a considerable variation from situation to situation that needs to be understood if risk estimation is to be useful. In transport, road travel is more risky than that by rail or air: on the roads, occupants of public service vehicles run one-seventh the risk of car occupants and one-160th the risk of motor cyclists (in terms of risk per distance driven). Users of urban roads run twice to three times the risk of users of motorways. Errors of human judgement, accentuated by alcohol, are much more important causative factors than either the road environment or vehicle defects. Similarly, at home it is the commonplace hazards, reflected in falls, scalds and burns, or mishaps with electrical services, that are the cause of most trouble rather than the rare but widely publicized building failure or gas explosion. Most fatal fires take place in the home, as a result of errors in everyday activities, and, as in travel, voluntary activities like smoking and drinking in inappropriate circumstances can accentuate the risk.

The British situation is closely comparable with that in other countries whose way of life is similar. Between 1970 and 1975, fire deaths in European countries appear to have ranged from 1 per 10^5 (Switzerland and Italy) to 2.4 per 10^5 (Finland), with the U.K., France, Denmark, Belgium, Sweden and Norway in the centre of the range (Wilmot 1979). In Canada and the U.S.A. the rates appear somewhat higher. Road casualty figures vary in a comparable manner: in the congested cities of developing countries they may, however, be up to six times as great as in the United States.

Risk evaluation brings us face to face with familiar divergences between probabilities and perceptions. In transport, the low-frequency high-mortality event is clearly perceived as especially repugnant, and investment to enhance safety on motorways, for example through the installation of crash barriers and lighting, has been given priority. So has investment to enhance the safety of air and rail travel, and to reduce fire risk in schools and other public buildings (even though schools, from the fire standpoint, are already outstandingly safe in Britain). There is nothing in this peculiar to the areas of transport and the built environment. As Golant & Burton (1969) (reported by Kates 1979) found, many of the hazards that are experienced least are the most feared. But the analysis does suggest that we face a similar dilemma. In statistical terms, the most efficient way to reduce risk may be to concentrate remedial action on the widespread, unspectacular event, and on the behaviour of individuals, rather than on further improving safety in those areas perceived as more hazardous than they are. But public concern cannot be ignored, and opportunities to reduce the risk of catastrophe must also be taken. In any case, the two classes form a continuous spectrum in several of the fields reviewed in this paper, and the best way of averting a low frequency, high mortality event on the roads or in fire (for example) is likely to be through applying the same kinds of measure likely to pay off in reducing more general public risk.

The views expressed in this paper are those of the author and not necessarily those of the Department of the Environment or the Department of Transport.

This paper could not have been prepared without the advice and guidance of colleagues in the Departments of the Environment and of Transport who have specialist knowledge of the fields with which it deals. Miss Barbara E. Sabey of the Transport and Road Research Laboratory, whose own research is referred to extensively, provided most of the information on risks in transport. Dr P. H. Thomas, of the Building Research Establishment's Fire Research Station, and Mr W. H. Ransom of the Building Integrity Division of B.R.E., provided details of risk from fire and from accidents in the home. Dr J. B. Menzies and Mr P. C. Roberts provided further contributions on buildings and on risk evaluation generally and on safety in rail travel. Miss Jean Irvine collated the information from these diverse sources. I gladly express my indebtedness to all these colleagues.

References (Holdgate)

Crossman, E. R. F. W., Zachary, W. B. & Pigman, W. 1977 *J. Fire* **71**, 67.

Department of Transport 1978 *Transport statistics, Great Britain 1968–1978*. London: H.M.S.O.

Department of Transportation 1978 *U.S. Department of Transportation News*, August. DOT 12278. Washington, D.C.

Golant, S. & Burton, I. 1969 *Avoidance-response to the risk environment*. Working Paper no. 6. Toronto: University Department of Geography.

Grattan, E. & Hobbs, J. A. 1980 *Transport and Road Research Laboratory report* no. LR 924.

Grime, G. 1979 *Transport and Road Research Laboratory report* no. SR 449.

Harland, W. A. & Woolley, W. D. 1979 *Building Research Establishment information report* no. 18/79.

Home Office 1978 *Fire Statistics U.K.* London: Home Office.

Home Office 1979 *Report of Her Majesty's Chief Inspector of Fire Services*. London: H.M.S.O.

Home Office 1980 *Review of fire policy. An examination of the deployment of resources to combat fire*. London: Home Office.

Kates, R. W. 1979 *Risk assessment of environmental hazard*. (SCOPE 8.) Chichester, New York, Brisbane and Toronto: J. Wiley.

McNaughton, I. K. A. 1980 *Railway accidents: Report on the safety record of railways in Great Britain*. Department of Transport. (Annual reports).

Office of Population, Census and Surveys 1978 *Mortality statistics, accidents and violence. Review of the Registrar General on deaths attributed to accidental and violent causes in England and Wales*. Series DH4, no. 5. London: H.M.S.O.

Sabey, B. E. 1980 *Transport and Road Research Laboratory supplementary report* no. SR 581.

Sabey, B. E. & Staughton, G. C. 1980 *Transport and Road Research Laboratory supplementary report* no. SR 616.

Sabey, B. E. & Taylor, H. 1980 *Transport and Road Research Laboratory supplementary report* no. SR 567.

Sheppard, D. 1975 *Transport and Road Research Laboratory supplementary report* no. SR 129.

Stratton, A. 1974 *J. Navig.* **27** (4), 407–449.

T.R.R.L. 1978 *Transport and Road Research Laboratory leaflet* no. LR 714.

Watts, G. R. & Quimby, A. R. 1980 *Transport and Road Research Laboratory report* no. LR 920.

Wilmot, R. T. D. 1979 *European Fire costs 1970–1975: a tentative analysis*. Geneva Association.

Discussion

J. P. LUMBERS (*Department of Civil Engineering, Imperial College of Science and Technology, London, U.K.*). An important area not already addressed, and one that would benefit from a risk analysis approach, is the setting and maintenance of water quality standards or objectives in natural water courses, both in the short and long term. The health hazards associated with the exceedance of various water quality standards need to be assessed in relation to water quality limits formulated on the basis of percentile non-exceedance values, which in turn should take account of the relative importance of the magnitude and duration of a water quality standard transgression. Potential applications of the technique also exist in the planning and design of water pollution control measures.

C. GREEN (*School of Architecture, Dundee, U.K.*). There is a danger of confusing perceived risk with preferences, of assuming that if the public demand a small risk to be made smaller then this must be because they overestimate the risk. This is to ignore the possibility that the public realizes that this is a small risk but, nevertheless, sees it as more important to reduce this risk rather than some larger risks.

When we presented respondents with a series of descriptions of supposedly actual dwelling fires, we found that the fires for which they wished to devote the most resources to reduce the risk of similar fires were not those that they saw as the most frequent fires. Instead, the most money was allocated to reducing those fires that were seen as resulting from the fault of those other than the occupants, even though these fires were seen as much less frequent than those seen as the fault of the house occupants. I would suspect that much the same thing happens with motorway cross-median accidents. We cannot assume that the public is only interested in minimizing the total risk in an activity.

Proc. R. Soc. Lond. A **376**, 167–179 (1981)
Printed in Great Britain

Risks and their control in civil engineering

By A. R. Flint
Flint & Neill Partnership, 1A Hobart Place, London SW1H 0HH, U.K.

The nature of risks in civil engineering, including those of collapse or unserviceability, is outlined. An indication is given of the orders of magnitude of failure probabilities based on experience and prediction. The attitudes of owners, the public and the profession to failures are discussed. Procedures used at present to combat these risks in design, construction and operation are summarized and their shortcomings discussed. The state of knowledge in reliability assessment is reviewed. Some of the difficulties in rationalizing the treatment of safety in this field are indicated, and suggestions made for action that could be taken to develop improved cost beneficial procedures.

Introduction

Civil engineering is a high-risk business. Its professional institution in Britain was founded 'to direct the great sources of power in Nature for the use and convenience of man'. Neither Nature nor man are readily predictable and it is therefore inevitable that the tasks fulfilling that aim are beset with hazard. Moreover, the design and construction of civil engineering works, without which the natural hazards of an uncivilized life would far exceed those experienced now, is undertaken within severe constraints.

The great utilities provided by civil engineering are almost always subject to limitations of time and budget. They are constructed for impatient clients, from imperfect materials and often by less-perfect humans. The profession is frequently at the forefront of knowledge, applying advanced technology to achieve the previously impossible without the benefit of gradual evolution. Seldom can a prototype be constructed and well tested before use: the products are usually 'one-off'.

Mistakes and failures cannot be hidden, and attract subjective public (and/or political) reaction out of proportion to their consequences. The increasing scale of engineering projects converts failures into catastrophes in the absence of a balanced view of risk. Progress in this field is hampered by a lack of a credible basis for assessing risks that may be anticipated, by the absence of a set of well defined safety criteria, and by the means to assess the benefits. Given these, designs could be optimized for maximum worth, and control procedures rationalized to achieve the best balance of expenditure on design, construction, inspection and maintenance.

This paper indicates the nature and orders of magnitude of risks in civil engineering, and discusses the progress made in some of the areas of uncertainty mentioned.

The nature of the risks

Risks to civil engineering projects fall into two categories: those of irreparable failure or collapse, and those of unserviceability when the utility fails to perform its function and requires repair other than normal maintenance. Associated with these risks are those to life and limb and to finances.

The first category includes hazards during construction as well as those in service. On average, more deaths and injury occur in the course of construction than in the later use of civil engineering works. In the United Kingdom there are, for example, about 120 fatal accidents annually in construction onshore (30 % of which are in civil engineering), representing an annual probability for the constructors of about 10^{-4} (Cotton 1978; H.M.S.O. 1978). Typical rates of accidents causing death or injury on construction sites in the U.S.A. and Australia have been of the order of 2×10^{-3} and 6×10^{-3} per annum (Cook 1975) respectively. There is a wide range of causes, the most dramatic being collapse of part of the works, commonly due to failure of temporary works. In the United Kingdom an average rate of death due to structural failure has been estimated to be only about 5×10^{-6} per annum (C.I.R.I.A. 1977), although 20 falsework failures involving injury were reported to H.M. Inspectors of Factories in 1974 (H.M.S.O. 1976) (about 2×10^{-3} times the work-force concerned with falsework).

An analysis of the failures of 143 bridges over a period of 130 years (Smith 1976) showed that 16 % occurred during construction, half of which were due to collapse of temporary works. There were on average two deaths per failure during construction.

Most casualties occur because of other bodily hazards connected with construction, including working at heights and falls of objects, handling goods, use of equipment, explosion, fire, earthfalls, and flooding. In addition, there are medical hazards affecting most branches of the industry. For those working under compressed air there are the risks due to decompression. Tunnellers may suffer from oxygen deprivation or the effects of fumes (Walder & McCallum 1975). Exposure to high noise levels is deleterious. Radiation risks are run by those using X-ray equipment for inspection. Dust is a universal hazard on building sites (I.C.E. 1971).

Table 1 shows the number and causes of accidents in building and engineering construction reported to H.M. Chief Inspector of Factories in 1978 (H.M.S.O. 1978). The causative factors assessed from a study of 250 accidents on structural works are summarized in table 2 (Brueton 1969).

The accident frequency rate in operation can exceed that in construction. For example, in sewage treatment a frequency of the order of 0.07 per annum has been reported (H.M.S.O. 1970).

Failures in service are usually of a structural nature. Their causes are diverse and

Table 1. Reported accidents in construction processes by type of accident, 1978

accident classification	building operations: fatal accidents	building operations: total reported accidents	works of engineering construction: fatal accidents	works of engineering construction: total reported accidents	all construction processes: fatal accidents	all construction processes: total reported accidents
falls of persons (excluding falls from vehicles)						
from heights	45	4510	11	481	56	5044
on the flat	—	3390	—	752	—	4198
total falls of persons	**45**	**7900**	**11**	**1233**	**56**	**9242**
falls of materials	12	1959	2	381	14	2363
excavations						
burial by fall of material	2	22	6	40	9	64
struck by material from the side (other than burial)	—	99	—	87	—	187
total excavation accidents	**2**	**121**	**6**	**127**	**9**	**251**
tunnelling						
burial by fall of material	—	—	—	—	—	—
struck by material from the side (other than burial)	—	1	—	5	—	6
total tunnelling accidents	—	**1**	—	**5**	—	**6**
lifting equipment						
hoists (excluding falls of persons and materials)	—	91	4	15	4	107
cranes and lifting machinery (including falls of persons and materials)	2	225	5	135	7	366
total lifting equipment accidents	**2**	**316**	**9**	**150**	**11**	**473**
machinery						
power and non-power machinery (other than lifting equipment)	1	786	—	359	1	1176
rail transport	—	8	2	29	2	37
non-rail transport						
vehicle in motion	6	737	6	414	12	1168
vehicle at rest	1	890	—	547	1	1460
total non-rail transport accidents	**7**	**1627**	**6**	**961**	**13**	**2628**
electricity	5	107	4	46	9	156
stepping on or striking against objects	—	2342	1	363	1	2723
hand tools (not power driven or cartridge operated)	—	1751	—	554	—	2323
poisoning and gassing	1	21	—	6	1	27
handling goods (not elsewhere specified)	—	5983	—	1649	—	7714
other accidents	—	2868	3	928	3	3861
total reported accidents	**75**	**25790**	**44**	**6791**	**120**	**32980**

TABLE 2. PERCENTAGE FREQUENCY OF CAUSATIVE FACTORS

inadequate supervision and control		47
unsafe means of access and place of work, including:		
working on open steelwork	4.8	
slippery or obstructed place of work	6.4	
muddy or untidy site	4.8	
traps in floor	3.2	
ladders, defective or unsecured	4.0	
		34.4
inadequate scaffolding or ladders		21
inadequate technical supervision		20
plant, equipment, fittings and details defective, overloaded or unsuitable for purpose		16.4
inadequate signals, instructions or briefing		14
inadequate inspection and maintenance		12
failure to provide and use suitable safety equipment, i.e. belts, harnesses, helmets, shoes, nets and goggles		11.5
failure to control suspended load from swinging or handling suspended load		8.8
insufficient or inadequate guys, props, shores or packings, or these not properly secured		6
poor or careless operation of crane		5.2
inadequate planning		5
inadequate coordination between contractors		4
inadequate labour force		4
inadequate slinging		3.6
high winds		3.6
failure to provide crane and sling hooks with safety catches		2.5
failure to isolate electric power lines in working area		2
inadequate lighting		2
failure to provide guards to dangerous machinery		1.2
fire and explosions		1.2
unsafe system		100

include the occurrence of unforeseen events, inadequacies in design, workmanship or materials and deterioration with time.

Over half the bridge failures in service appear to have resulted from flood and foundation movement (Smith 1976), and 10 % were due to collision from shipping (this latter hazard has recently been seen to be very serious for bridges with structure adjacent to waterways). Less than 10 % could be attributed to inadequate design against loadings conventionally catered for. It has been estimated that the rate of failure of suspension bridges within their normally expected lifespan prior to 1968 has been of the order of 1 in 14. The corresponding rates for other major unconventional bridges, ordinary road or railway bridges and small-beam bridges have been 10^{-1}, 10^{-2} to 10^{-3} and 10^{-3} to 10^{-4} respectively (Pugsley 1968). The number of people at risk at the time of failure of a bridge in service is difficult to assess as can be judged from the analysis of past events (Smith 1976). Deaths resulting from a failure have ranged from none (as, for example, with the ill-fated and sensational Tacoma Narrows bridge) to 75 (for the equally historic first Tay bridge). The average appears to have been about 3 per failure.

Tall towers and masts are at risk under extreme environmental conditions, and failures have occurred due to high wind and due to the effects of icing. Experience with transmission line towers in the U.K. has shown an average failure rate of about 10^{-5} per annum in a recent 15 year period, the predominant causes being extreme icing on conductors or faulty foundations.

Broadcasting towers have had an annual failure rate of the order of 10^{-4} and guyed masts about 10^{-3} due to all causes.

Some have collapsed under strong winds and others due to severe vibrations caused by aerodynamic stability. There have been a number of instances of impact from aircraft on masts; at least one transmission tower was felled by such a collision.

Also prone to the hazards of environmental forces are offshore platforms. There have been a number of failures of these in storms in various parts of the world, one recently in the North Sea.

Extreme cyclones and tornadoes present a hazard. For example, in the worst zone in the U.S.A. the annual risk of an occurrence of a tornado per square mile is about 10^{-3}.

Large dams have shown a relatively high failure rate. Of those completed before 1965, about 1 % of arch and earth-filled dams, 2 % of buttress and rock-filled dams and 0.4 % of gravity dams had failed (International Commission on Large Dams 1973). The failures resulted mainly from a loss of stability, deterioration or overtopping. No general statistics of consequences of failure have been obtained, although it has been estimated that 2×10^{-5} of total deaths per year in the U.S.A. may have resulted from collapses of dams (Buehler 1975).

A study being undertaken of the performance of dams and reservoirs in the U.K. (A. I. B. Moffat, personal communication) has shown that some 42 out of a population of about 5000 reservoirs (0.8 %) have failed and that there have been a larger number of occasions on which failure has been narrowly averted by lowering water levels or by repair. The average age of the reservoirs is about 95 years and many of the smaller of these are unattended. With a noted tendency for failure rate to increase with age and with no clear responsibility for inspection and maintenance of the small dams not covered by the Reservoirs (Safety Provisions) Act, 1930, there is likelihood of an increase in failure rate in the future.

The last failure in the U.K. causing loss of life appears to be one that occurred in 1930.

Although no statistics relating to frequency of tunnelling collapses appear to be available, there have been numerous mishaps, the most common of which have resulted from wash-out from water-bearing strata (Muir Wood 1975).

Attitudes to risks in civil engineering

Professional education in the past has seldom included any instruction in matters of safety, and there has been an unjustifiable confidence that absolute reliability can be achieved. This confidence has been shared by the lay public, despite the lesson of history. As a result, failures in civil engineering have commonly been treated as disasters often out of proportion to the cost of life.

The reaction of the media and politicians to mishaps causing death has tended to be emotional, with the consequence that the profession has been forced to take precipitate action to 'bolt the stable door' on several recent occasions. This has been particularly noticeable in the building field.

Public enquiries and commissions have served an important role in establishing causes, but their aftermath has tended to bring hasty changes in practice and increased bureaucracy without due consideration of the economic implications.

In recent years there has been a radical change of attitude of engineers and owners of civil engineering works. The profession has shown great interest in attempts to rationalize safety measures and there have been changes in the philosophy of design. The concept of choosing reliability levels related to social and economic consequences of failure is becoming respectable (Bridle 1975). The need to aim for uniformity of reliability within utilities serving the same purpose is widely accepted (*Journal of the Institution of Structural Engineers* 1979). Attention is being paid to control of risks arising from gross errors and poor communications (Flint & Quinion 1978). The statistical definition of design criteria has increased the awareness of natural uncertainties. Safety at work sites has been a matter for considerable discussion (I.C.E. 1969). There are now serious attempts to collect and analyse information concerning failures and unserviceability.

There are impediments to the acceptance of risks at economic levels. There is no benefit to the consulting engineer in reducing safety margins and hence the cost of the works. His practice will suffer seriously in the event of a failure disproportionately to the fee income for the commission concerned. His fee, if based on the cost of the works, will be reduced. The risk of a claim of negligence will increase.

The benefits lie with owners, particularly those of large numbers of projects, and it is to them that we must look for a lead in balancing risk against cost, including the cost of design and error control. They in turn are inhibited by public and political attitudes. It is to be hoped that following the discussion there will be action to stimulate public interest in putting risk into perspective and to curb irresponsible political alarms.

Combating risks

Awareness of risks

The civil engineering profession is constantly aware of risks associated with the construction and operation of works although, as a result of litigation, it has often been ill-informed of the frequency and causes of failures. Technical news magazines frequently provide the only source of information concerning events. There have been few authoritative published reviews of collapses and almost none on cases of unserviceability.

There have been a number of notable reports by Royal Commissions and Committees of Inquiry set up after major events which have greatly assisted the understanding of both the nature of risks and their causes. These have commonly led to changes in practice and other action to increase awareness: for example, the Report of the Royal Commission of Inquiry into the collapse of the Second Narrows bridge, Vancouver (1958), during construction led to changes in conditions of contract and clearer responsibilities in relation to temporary works. Failure, for similar reasons, of the Loddon bridge was followed by a report of an Advisory Committee (H.M.S.O. 1976) and a conference at the Institution of Civil Engineers about falsework hazards. The Ronan Point building collapse and the high-alumina cement incidents led to various publications by the Institution of Structural Engineers (1976). All such post-accident actions are generally aimed at avoiding repetition of types and causes of failure.

The Standing Committee on Structural Safety set up by the Institutions of Civil, Municipal and Structural Engineers in 1976 has as its terms of reference to study trends in the structural field and to consider where work or warning of risk is desirable from the standpoint of safety. In its reports it reviews potential hazards that could in the future cause failure. As a focus for fears or clairvoyancy in the profession, this recent innovation should alert engineers to possible risks.

The Department of Energy Petroleum Engineering Directorate has likewise formed advisory groups to consider where research or development are desirable to reduce the risks to offshore installations, and these groups endeavour to foresee future hazards.

Design risk control

Although relatively few reported failures have stemmed from oversight in design, investigations into some important events have indicated the importance of controlling these. For example, the Committee of Inquiry into the failures of box girder bridges found widely varying standards of design and checking for bridgeworks (H.M.S.O. 1973).

The Department of Transport now requires that the designs of all their bridges of any importance are approved in principle by their Technical Approval Authority. For the major bridges, certification of designs is provided by a checking team independent of the original designers (Bridle 1975).

Similar checking procedures for offshore installations were adopted after the Offshore Installations (Construction and Survey) Regulations, 1974. Certification authorities were nominated who undertake design appraisals before certification of structural adequacy.

The design of large reservoirs and dams in the U.K. is required to be undertaken under the direction of a member of a special panel set up under the Reservoirs Act,

1930, the member being personally responsible. A similar provision is made in the 1975 Reservoirs Act, although this remains to be implemented.

For structures the designs of which are prepared by contractors, such as towers and masts, it is common for owners to call for type tests (as for transmission towers) or to arrange for design appraisals by consulting engineers.

Although the cost-effectiveness of such procedures cannot at present be measured, experience has already shown that they detect errors in principle and mistakes in detail.

Designs for civil engineering works of a structural nature and for earthworks are commonly based on Standards and Codes of Practice published by the British Standards Institution and drafted by the profession. These frequently prescribe the loadings to be assumed in design as well as giving rules for calculating strengths. Safety margins specified have usually been judged by reference to previous practice and are not directly related to any target risk levels or based on adverse experience.

Structural design 'safety factors' are primarily intended to allow for uncertainties in loading and strength, and cannot effectively combat risks due to gross errors in design and construction. Analysis of failures suggests that current practice in civil engineering uses factors that maintain levels of risk in the absence of blunders to very low levels (the fraction of failures attributable to inadequate strength to sustain predictable loads is less than 10 %) (C.I.R.I.A. 1977). Indeed, it is probable that it would in many cases be cost-beneficial to reduce the design margins of safety and increase expenditure on other forms of control. Typically a reduction in safety factor (and material cost) by the order of 10 % increases the theoretical failure probability tenfold. With very low values of probability this may not result in a significant number of failures. The saving of perhaps 3 % in overall construction cost devoted to other forms of control could greatly increase reliability; for example, an independent design check costs of the order of 1–2 % of the cost of the works.

Codes provide some 'norm' for design practice but tend to be inflexible and conceal the uncertainties in design. Furthermore they may, by their incompleteness, lull the designer into overloading certain hazards: for example, the Ronan Point flat failure due to explosion could in part be attributed to the relevant Codes (deemed to satisfy the Building Regulations) failing to prescribe design criteria for such a risk. The collapse of the cooling towers at Ferrybridge (I.C.E. 1967) would have been averted had the current codes properly considered the discontinuity in the relation between wind load and stress in the reinforcement under strong winds. Codes by themselves are no substitute for independent thought.

There are also risks of incorrect interpretation. An 'examination' set by the BS 5400 committee on fatigue in bridges to test the application of draft rules showed a remarkable range of calculated fatigue lives for set examples. This demonstrated the hazards associated with set rules for a complex subject.

Alternative aids for design have been produced in the form of guidance notes. The Department of Energy has published such guidance in a form that sets general criteria for design of offshore structures but which allows the designer to make use of the best data available for assessing structural response to extreme environmental conditions. A number of design guides for specific problems have been produced by the Construction Industry Research and Information Association. There are also condensates of existing technical knowledge, published by the Engineering Sciences Data Unit, which provide designers with carefully sifted information.

Construction risk control

Although much has been written about the nature of risks in construction and their control (I.C.E. 1971), there is no common standard of safety. Quality of materials and workmanship is commonly specified and supervised by the Engineer for major works and it has been unusual for material deficiencies to contribute to failures. Faulty workmanship or materials should be detected by normal inspection procedures. However, for tasks involving special skills and experience, such as tunnelling, much still depends on the contractor's personnel.

Inadequate supervision and unsatisfactory temporary works have been identified as major factors influencing risk. Ignorance and carelessness play an important part (H.M.S.O. 1976; Smith 1976; Brueton 1969; Matousek 1977; Melchers 1978). All these are controllable, given adequate resources and organization.

Hazards at the interface

Risks exist at the interface between design and construction that deserve some control (Flint & Quinion 1978). Communication of design intentions from designers to contractors and of proposals for temporary support and for other construction conditions in the reverse direction has been shown to be important. The collapse of the Yarra bridge resulted in part from shortcomings in communication between the designer and the builder (Government of the State of Victoria 1971). That of the Vancouver bridge could have been avoided had the engineer checked the scheme for temporary support.

These hazards are affected by human relationships, use of language, experience of personnel and finance. They can only be controlled by employment of competent, compatible and literate staff on both sides without undue financial or temporal constraint on their work. In an attempt to reduce the risks, the Institution of Structural Engineers (1975) published general guidance on communication of structural design.

Control of loading

Certain types of works may be safeguarded by control of loading. For example, traffic control can be used on bridges, and tethered offshore platforms and tower cranes can be positioned to minimize the effects of extreme storms. These forms of control can reduce risks.

Prediction of accident-proneness

The history of failures has shown that there are certain influences on the safety of works that should be considered before the works are executed. The use of new or unusual materials, types of structure, or methods of construction, the state of knowledge, the industrial, financial or political climates and the capabilities and organization of the design and construction teams have been identified as significant (Pugsley 1973). Although there does not yet appear to be any recognized procedure to allow for these, logical procedures for employing subjective judgement in rating accident-proneness are being developed (Blockley 1980).

Reliability assessment

The reliability of civil engineering works (or, conversely, their probability of failure in a given period) depends on the characteristics of the phenomena that they have to resist and on their resistance. In many cases the phenomena are natural and random, such as flooding, wind and wave; in others they are man-made and random, such as highway loading; in yet others they are controllable or determinate, such as water loading in a water tank. Resistance is also commonly random owing to variability in material and workmanship and owing to the material quality, quantity and disposition specified by the designer.

The uncertainties in performance therefore consist partly of physical uncertainties, beyond the control of man but often capable of statistical description, which are subject to error. There are also the uncertainties concerning the relations between the basic variables, such as wind speed and load, and those between load and load effect. These uncertainties in knowledge can only be judged and not measured. The tails of statistical distributions are particularly difficult to establish.

Despite these problems, statistical reliability theory is being widely used in civil engineering. It has for some time been the practice to express design criteria in terms of events having a prescribed probability of occurrence in the design life of the system. Offshore platforms in the North Sea are designed against the 100 years return period sea state, the coincident wave heights and periods being predicted by statistical extrapolation from records. Design wind speeds for mast structures are chosen to have return periods appropriate to their importance, commonly 30–50 years. The new Thames barrier and the flood defences associated with it are designed for a 1000 years return period flood with land levels estimated for the year 2030 (G.L.C. 1969–70). (The flood risks before bank raising in the early 1970s were of the order of 1 in 10 per annum and are at present about 1 in 55.) Dams and reservoir spillway capacities are based on statistical assessments of extreme flooding.

In earthquake zones, designs are based on accelerations and velocities of ground movement predicted to have probabilities of not occurring in the design life commensurate with the importance of the works (commonly 90 %).

Similarly, extreme man-made events are being assessed statistically. For example, extreme loads for highway bridges are calculated from observations of vehicle weights, traffic patterns and accidents. Design loads are intended to represent events with a 5% probability of occurrence in a bridge life of 120 years.

The probabilities of failure associated with these design criteria are far less than those of occurrence of the prescribed events as a result of the use of safety factors. Moreover, design resistance of structures is usually taken as the 95% fractile below which only 5% of all strengths should fall. Combining the statistics of loading and strength, with due allowance for modelling uncertainty by means of reliability theory, allows estimates to be obtained of conditional failure probabilities of structures. The probabilities are conditional on the validity of the modelling uncertainty assumed and on the risk's being unaffected by matters not represented by the statistical model.

As examples of the results of such analysis, table 3 shows the orders of annual failure probability of several types of structures designed to British practice, based on assessment of component reliability.

TABLE 3. TYPICAL CONDITIONAL ANNUAL FAILURE PROBABILITIES, P_f, FOR ENGINEERING STRUCTURES

	P_f	references
reinforced concrete and steel buildings	10^{-6} to 10^{-14}	C.I.R.I.A. (1977)
steel towers	10^{-2} to 10^{-4}	B.S.I. (1978)
steel and composite highway bridges	10^{-5} to 10^{-13}	Flint & Neill/Imperial College (1980)
offshore jacket structures	10^{-2} to 10^{-5}	Fjeld (1977), Marshall (1969)

PROBLEMS IN RATIONALIZING APPROACHES TO SAFETY

Optimum designs best suited to all relevant constraints are achieved in Nature by evolution. In the ages of slow development, the same processes were to a considerable extent the basis of engineering, and remain in the construction of repetitive artefacts. Evolution involves large numbers of failures. The civil engineer concerned with individual projects using new techniques does not usually have the benefit of positive experience and is expected to be less profligate with failures than Nature. He therefore faces the problem of setting and achieving target levels of risk.

Acceptable risk levels remain to be defined. They can be evaluated on purely economic grounds provided that the consequences of failure can be estimated. Economic optimization procedures are available for treating those risks that are within the control of design safety margins, and frequently lead to target safety factors and failure probabilities much higher than those in existing practice (C.I.R.I.A. 1977; Bridle 1975).

Less easy to define are the social criteria for risk control. These have been extensively studied but there is no quantitative set of rules to be followed (C.I.R.I.A. 1977).

As mentioned earlier, only a small fraction of total risk is controllable by design. The large remainder (grouped as dominated by error or accident) at present defies mathematical modelling since, strangely, there has been little research into the statistics. In consequence, the direct use of reliability theory is not sufficient to indicate safety provisions to meet target risk levels, even if these can be defined.

As a short-term expedient, calibration against previous practice, using reliability theory as a comparative tool, is being used to derive design safety factors (C.I.R.I.A. 1977; Flint & Neill/Imperial College 1980). The aim is to produce greater consistency in reliability without altering current average risk levels. To do this there is a need to improve knowledge of the physical and modelling uncertainties and to place more reliance on computed failure risk.

Of greater difficulty is the quantitative assessment of the risk of blunders. Work on risk analysis by using fault-tree methods or other techniques developed for quality control in production industries is of limited value in judging the cost effectiveness of control in civil engineering. What are required are extensive studies of causes of failures and of alternative control procedures that could have prevented them. Reviews of the outcome of present checking and supervision methods could also indicate the benefits that they provide, and the associated costs.

Above all is the need to define what is 'rational'. It is unlikely that we are directing or controlling the great sources of power in Nature to the maximum benefit to man with the minimum consumption of the world's resources. It is to be hoped that these discussions will promote interest in establishing the definition.

References (Flint)

Blockley, D. 1980 *The nature of structural design and safety*. Chichester: Horwood.

Bridle, R. J. 1975 In *hazards in tunnelling and on falsework*. London: Institution of Civil Engineers.

Brueton, G. 1969 In *Safety on construction sites*. London: Institution of Civil Engineers.

B.S.I. 1978 *Draft Code of Practice. Lattice towers: loading; Commentary*. London: British Standards Instition.

Buehler, B. 1975 In *Responsibility and liability of public and private interests on dams*. American Society of Civil Engineers.

C.I.R.I.A. 1977 *Rationalisation of safety and serviceability factors in structural codes*. Report no. 63, Construction Industry Research and Information Association.

Cook, G. P. 1975 In *Hazards in tunnelling and on falsework*. London: Institution of Civil Engineers.

Cotton, R. M. W. 1978 *Proc. Instn civ. Engrs* (1) **64**, 181–183.

Fjeld, S. J. 1977 In *Offshore Technology Conference*.

Flint & Neill Partnership/Imperial College 1980 Derivation of Safety Factors for BS 5400, Part 3. Report for Department of Transport. (Unpublished.)

Flint, A. R. & Quinion, D. W. 1978 *Proc. Instn civ. Engrs* (1) **64**, 529–532.

Government of the State of Victoria 1971 *Report of the Royal Commission into the failure of West Gate Bridge*.

G.L.C. 1969–70 *Thames Flood Prevention. Thames Barrier Project:* report of studies.

H.M.S.O. 1970 *Report of Working Party on Sewage Disposal: Ministry of Housing and Local Government*. London: H.M.S.O.

H.M.S.O. 1973 *Report of Committee: inquiry into the basis of design and method or erection of steel-box girder bridges*. London: H.M.S.O.

H.M.S.O. 1976 *Final Report, Advisory Committee on Falsework*. London: H.M.S.O.

H.M.S.O. 1978 *Health and Safety, manufacturing and service industries*. London: H.M.S.O.

I.C.E. 1967 *Natural draft cooling towers – Ferrybridge and after*. London: Institution of Civil Engineers.

I.C.E. 1969 *Safety on construction sites*. London: Institution of Civil Engineers.

I.C.E. 1971 *Hazards in construction*. London: Institution of Civil Engineers.

Institution of Structural Engineers 1975 *Communication of structural design*.

Institution of Structural Engineers 1976 *Criteria for structural adequacy of buildings. Journal of the Institution of Structural Engineers* 1979 A**59**, no. 9.

International Commission on Large Dams 1973 *Lessons from dam incidents*.

Marshall, P. W. 1969 *Proc. Am. Soc. civ. Engrs J. struct. Div.* ST**12**, 2907–2929.

Matousek, M. 1977 In *Colloquium on Inspection and Quality Control*, Cambridge.

Melchers, R. E. 1978 *Proc. Instn civ. Engrs* (2) **65**, 791–807.

Muir Wood, A. M. 1975 In *Hazards in tunnelling and on falsework*. London: Institution of Civil Engineers.

Pugsley, A. G. 1968 *Struct. Engr* **46** (7), 197–201.

Pugsley, A. G. 1973 *Struct. Engr* **51** (6), 195–196.

Smith, D. W. 1976 *Proc. Instn civ. Engrs* (1) **60**, 367–382.

Walder, D. N. & McCallum, R. I. 1975 In *Hazards in tunnelling and falsework*. London: Institution of Civil Engineers.

Proc. R. Soc. Lond. A **376**, 181–192 (1981)
Printed in Great Britain

OVERVIEW

Risk assessment: use and misuse

BY D. W. PEARCE
University of Aberdeen, Edward Wright Building, Dunbar Street, Aberdeen AB9 2TY, U.K.

Risk analysis has emerged in recent years as a technique that allegedly aids decision-making in respect of the choice between alternative investments, especially energy investments. Once it is recognized that the term 'risk' has at least two characteristics – the probability of the event's occurrence and its associated magnitude if it does occur – the perception of risk as a composite entity becomes important. This perception must be translated into some formulation of the way in which individuals 'trade-off' probability and magnitude if risk analysis is to have a role in decision-making. But this in turn entails the requirement that we know the nature of this trade-off (the utility function) and how to aggregate from individuals to society. Even where probabilities can be established, perceptions of the magnitude of the event will differ. It is therefore not possible to standardize the outcomes to which probabilities are attached in the manner adopted by most risk analyses. The paper argues that the only proper approach is through the search for the indirect revelation of individuals' preference for both the risks of the projects in question and their associated benefits. Formulated in this way, risk analysis becomes a sub-category of cost–benefit analysis and is not in itself a means of evaluating projects.

RISK AND ASSESSMENT

Any corporation, public utility or government will react to criticism of its activities by seeking what are ostensibly new ways to further the acceptable image of their activities. The recent and continuing fashion for engaging in risk analysis is no exception. Nor should it occasion surprise that the energy industries have been the focus for risk analysis since the debate over which 'energy path' to take in the next few decades shows no signs of abating, and the constituent industries in each of the alternative paths have various mixes of risk characteristics. In particular, conventional futures seem bound to contain low-probability, high-magnitude social costs. Liquefied gas terminals and nuclear power are the obvious examples.

Unfortunately, while risk analysis has emerged as a defensive posture and is extensively developed and analysed, it is none too clear that the participants to the debate agree on what they mean by the term 'risk'. More importantly, and this is the subject of this paper, inadequate attention has been paid to the issue of what risk analysis, whatever it means, has to do with decision-making. While it has its uses, the procedure of looking at risks associated with a given economic activity is now wrongly advocated in certain quarters as a decision-making calculus in itself. That is, it is argued either that it acts as an 'input' to the wider process of decision-making, or that it can actually guide decision-makers in choosing between investments.

I shall attempt to show that risk assessment, properly interpreted, is nothing new and that the manner in which it can be incorporated into decision procedures has long been known. It does not follow from such a statement, if it can be proved, that the activity of thinking about risks, drawing up risk 'league tables' or whatever, is meaningless. It does mean that risk assessment is not a separate technique indicating anything at all about the social desirability of a given activity or investment.

In order to proceed to the stated goal, we must of necessity make some heroic assumptions. We shall first assume that there is an agreed definition of risk that we take to be a two-dimensional entity comprising (*a*) the probability of an event's occurrence, and (*b*) the magnitude of that event. The review by Schaeffer (1978) suggests such a definition would command general though not unanimous consensus. Secondly, we shall assume that whoever assesses the risk perceives correctly its nature and magnitude. We therefore postulate a set of rational risk-perceiving individuals. That complexities abound when we ask whether individuals correctly assess risk is well documented in the work of Fischoff (1979). Some other assumptions will be made clear as we proceed, but one axiom should perhaps be made clear at the outset. We shall take it that it is the preferences of individuals as they relate to perceived risk that count in the establishment of decision-rules. We might dub this an 'axiom of naïve democracy'. It is familiar to economists in the neoclassical tradition as the underlying value judgement of normative economics. In itself it is a morally contentious axiom as the literature testifies (see Nash *et al.* (1975) and Rowley & Peacock (1975)), but one that we retain.

Trading-off probability and magnitude

The first task is to establish the way in which our hypothesized rational individual will substitute between negative outcomes and the probabilities that those outcomes will occur. We confine ourselves to negative outcomes (disutilities, suffering) even though we should not forget, as most risk analyses seem to, that benefits are also probabilistic. Concorde is an example of an investment that had a high monetary cost that finally went well beyond the best guesses of the experts as to the upper bound of probability, and a flow of benefits that went well below the lower bounds of expected probability (Henderson 1977).

Figure 1 suggests the structure of an individual's preference function as it relates to the dual constituents of risk. The curves show combinations of probability of occurrence and the magnitude of the event between which the individual is indifferent. Observing curve I_1 it will be seen that the individual is assumed to be indifferent between some combination of probability and magnitude (p_A, M_A) at A and some other combination (p_B, M_B) at B. The rationale here is simply that the individual will 'trade' at a lower probability (at B) for a higher negative outcome, and vice versa. The concavity of the curve I_1 is established by arguing that greater and greater decreases in the magnitude of the negative outcome will be necessary to compensate the individual for higher and higher probabilities that the negative outcome will occur, as is seen by moving from B to A.

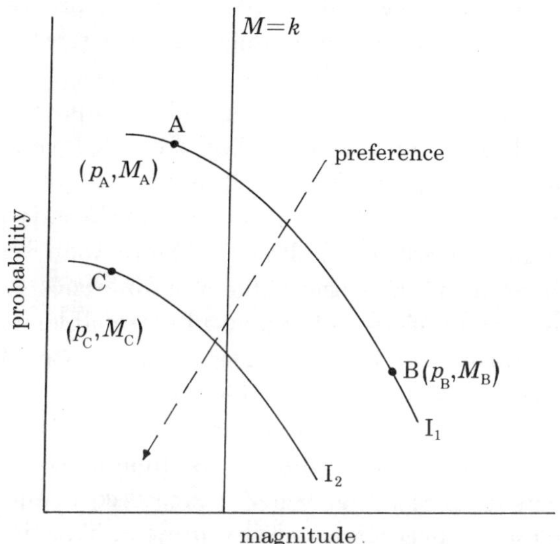

FIGURE 1. Preference functions for risk.

Clearly, there is a whole 'family' of such curves and curve I_2 illustrates one other. Any point on I_2 is superior to any point on I_1 in the sense that it is preferred to any point on I_1. This is obvious for, say, points C and A since A has both a lower probability of a negative outcome and a smaller negative outcome compared with A. C may therefore be said to 'dominate' A. The fact that C is preferred to B is less obvious. No proof is given here but it is in fact no different from the proofs given by economists for the existence of so-called 'utility functions', which is essentially what figure 1 illustrates. For formal proofs see Green (1976), and for an indirect demonstration see Ryan & Pearce (1977).

From figure 1 we can derive certain conclusions. First, if we are comparing any two projects such that one has a lower probability of negative outcomes *and* a lower magnitude of outcome, that project will be *partially ranked* above the one with higher values. The ranking is partial because, as we shall see, many other factors besides risk must enter any decision-making calculus. This ranking is illustrated by the superiority of project C over project A in figure 1. We require no knowledge of the utility function – the shape of the curves and their relationship to each other – to establish this somewhat trivial findings. Secondly, provided we *do* know the utility function, and not otherwise, we can rank projects where complete vector inequality does not apply, i.e. projects such as C and B where one component of risk is higher and the other lower. Note that observations about probability alone tell us absolutely nothing. We require knowledge of the individual's utility function, and that we take to be part of the problem of discovering how risk is *perceived*. Thirdly, we can observe that there is a special case of 'dominance', which permits a statement of preference to be made. This occurs where the two projects in question have the same magnitude of outcome, say $M = k$ in figure 1. Here it is obvious that the project with the lower probability dominates. Indeed, we simply have a special case of dominance and we require no knowledge of the utility function.

The caveats must also be made clear, even for such a simple construct. We have assumed that individuals are risk averse; if they are risk lovers the direction of preference will be the opposite to that shown in figure 1. To bring the analysis a little closer to decision-making we also have to translate the curves in figure 1 into *social* preference functions. Here we shall do no more than assume that some aggregation of the curves in figure 1 (individuals' indifference curves) is possible without offending one or more of a set of axioms of reasonableness. That this may not be possible is the subject of an extensive literature initiated mainly by Arrow (1963) and well surveyed in Mueller (1979). Lastly, even if we can derive some ranking from figure 1 it tells us nothing at all about the desirability or otherwise of the projects. For, even in the cases where 'risk dominance' is established, we have not been able to demonstrate that benefits exceed costs and we take that as a formal requirement for the acceptance of any project. That is why we refer to any derived ranking as a *partial ranking*: we require far more information than any risk assessment has so far provided.

In terms of figure 1, therefore, we can establish that activities of risk comparison such as that carried out by Inhaber (1978) and Holdren *et al.* (1979) – each with rather opposed findings – will have validity if and only if they can assume that they have properly defined risk in terms of risk per unit of constant negative outcome. That is, they must be comparing points on a line such as $M = k$ in figure 1. If, for any reason, those negative outcomes are not the same, we require a knowledge of individuals' utility functions (and a satisfactory rule for the aggregation of preferences). To illustrate why this is important, we have to dwell briefly on what the 'magnitude' of an event means. One is tempted to accept the view that 'a death is a death is a death', however caused. In reality we know that is not the case. Even assuming, as we have done, some concept of perfect knowledge and foresight, some ability to empathize with the victim of suffering, it remains true that a sudden death in a coal-mine explosion is most unlikely to be perceived in the same way by the sufferer as death over a period of years from pneumoconiosis or cancer from an excess dose of radiation. Nor is the death of a 6 year old child the same as a death of a man of 89, even if both die the same way. Nor is the death the same 'event' if it occurs through voluntary action, compared with actions that one may well have implicitly voted for but which nonetheless is perceived as being 'imposed' by some entity whose bureaucracy lies beyond public control. The list of differentiating features is endless and they are well rehearsed in the literature. The point for our purposes is *not* to say that one risk analysis is 'better' than another. Rather it is to raise the question of whether such risk analyses have much use at all. We return to the policy implications shortly. For the moment we observe the absolute necessity of assessing utility functions. Note that this requirement does not disappear if we substitute expert assessment for individuals' preferences. We must still be able to rank across events without vector inequality and that, as we have seen, still requires knowledge of the utility function. If any doubt remains on this issue, the reader is invited to draw shallower indifference curves that pass through C and B in figure 1. It is a simple matter to secure a context in which B is then preferred to C.

Benefits and acceptable risk

We can now attempt to introduce benefits directly into the picture. The rationale for doing this is that we only have a partial ranking so far. We have ignored all other costs and all benefits. Apart from the fact that the calculus is incomplete without reference to benefits – no one chooses between projects on the basis of risks alone – we argue that the concept of an 'acceptable risk' is without proper meaning unless benefits are known. Again, this is not a widely accepted definition of acceptable risk. It is used, for example, in the work of Starr (1969, 1979; Starr *et al.* 1976) and stands in somewhat stark contrast to the implied definition in Rothschild (1978). The latter would relate the concept of acceptability to other risks implicitly accepted by individuals in their lives. There are, however, rather severe problems in determining whether the term 'acceptable' is proper for the so-called 'baseline' risks, such as risk of death in a road accident, the risk of death from viral disease and so on. Apart from the facts that the probabilities in question vary so that it is unclear *which* baseline risk estimate is the one being worked to, we have already observed that the nature of the risk is important. Additionally, risks must not be cumulative. Adding the risk of death from some energy conversion process to the risk of death from a road accident ignores the increasing disutility that almost certainly attaches to the creation of extra risk. Finally, acceptable risk on this definition ignores the cost of reducing risks to the baseline level. If the cost of reducing such risks differs, we can expect differing perceptions of their acceptability to arise.

In any event, rationality of the risk-averse stereotype that we have created would be incomplete unless risks were related to benefits. This rationality has its foundations in the most casual observations about gambling and the probabilities of securing low and high stakes.

However, to integrate our risk-indifference curve approach with benefits requires us to 'collapse' the two dimensions of risk into one. This we can do by observing figure 2, in which it is assumed that society faces some risk levels, none of which is zero. However, from the utility function in figure 1 it can be inferred that there are combinations of probability (p_e) and magnitude(M_e) that are equal in utility terms to some magnitude ($\bar{M}$), which is certain ($p_e = 1$). $\bar{M}$ is the *certainty equivalent* level of risk. Figure 3 then shows the level of disutility (suffering) measured against various certainty equivalent levels of risk. Since we have reason to suppose that the disutility increases at an increasing rate, the disutility curve in figure 3 has the shape shown.

We need to make one more transformation of the disutility curve before introducing benefits. Typically, benefits will be expressible in terms of money values, pence per kilowatt hour or whatever. It follows that we need a money valuation of risk, otherwise we have no basis for comparison. Once again the matter is a

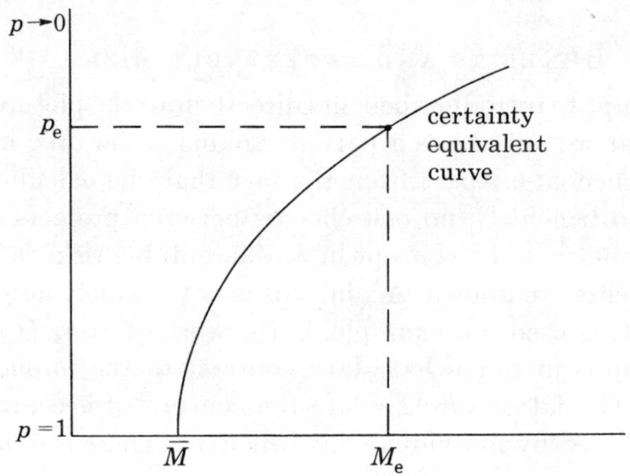

FIGURE 2. Risk certainty equivalence.

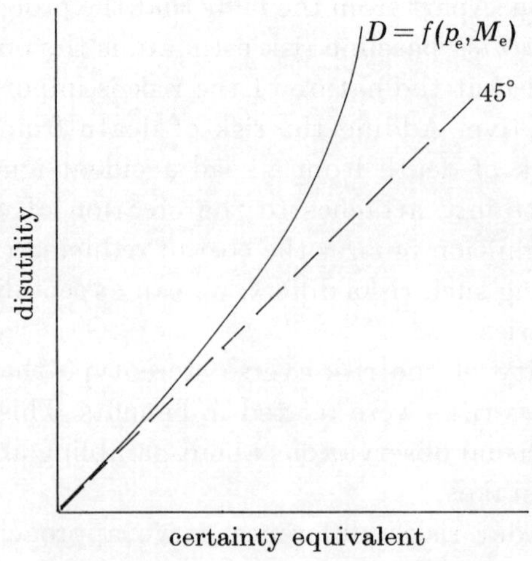

FIGURE 3. Disutility and risk.

controversial one since it involves the placing of money values on precisely those issues that so many people feel are 'beyond price'. Quite what this instinctive reaction to money valuation means is not always clear. It presumably does not mean that life has infinite value. That is inconsistent with actual behaviour, although entirely consistent with the sum required in compensation for *certain* death. It more likely means that some people see the attachment of money values

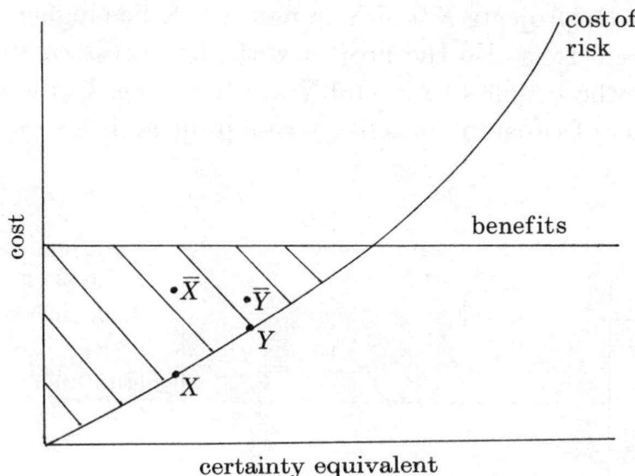

FIGURE 4. The basic cost–risk–benefit calculus.

to human life and suffering as in some way morally offensive. Yet any decision implies a value of human life and that alone is sufficient for us to justify converting the disutility function into a cost-of-risk function, using whatever values are separately determined to apply. For a discussion on approaches to valuing suffering and human life, see Mooney (1977). The money valuation procedure may leave the general shape of the disutility function unaffected. It may actually increase its curvature (steepness) if we discover that higher units of disutility attract higher money 'prices'. It can of course also widen the variance around the resulting cost function.

In figure 4 the disutility function is repeated, but this time converted to money terms. Also shown is a benefit function which has been taken to be invariant with the level of expected value of risk but which could take on various shapes. This potential variability does not affect our analysis. On the benefit cost rule, we deem any project with benefits greater than costs to be acceptable. If the costs of the risk were the *only* costs, then any project falling within the shaded area would be 'acceptable'. But if this is the final calculus to be used we can see immediately that it is nothing else but cost–benefit analysis with the only costs simply expressed as the money value of the risks involved. To this extent, what began as 'risk analysis' is seen to be a special case of cost–benefit analysis. Economists are not at all unused to integrating risk and uncertainty into such analyses, although all of the problems that we have come across, notably the discovery of the (dis)utility function relating to risk, are still applicable. Note that the *only* circumstances in which simple comparisons of *probabilities* of death or suffering yield the same results as cost-benefit analysis are those in which (i) the magnitude of the risk event is identical between the alternatives in question and (ii) the benefits are identical. I have suggested already that requirement (i) is almost certain not to be met, in which case we require knowledge of the cost function in figure 4. If, however, risk analysis is transposed to *utility of risk* or *cost of risk*, then the results are the same. Consider projects X and Y in figure 4. X has higher net benefits and is therefore preferred. It is also the project with the lowest cost of risk. But this result holds only if the benefits to X and Y are the same. I now proceed to show that the assumption of constant benefits across projects is also unrealistic.

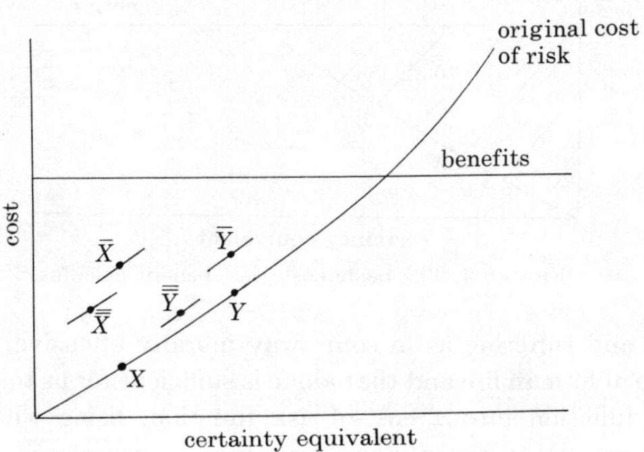

FIGURE 5. Effects of risk reducing programme.

THE REMAINING INPUTS

It is a relatively simple matter to modify figure 4 to allow for most other costs. The costs of constructing plant, for example, simply means adding to the cost of risk. If project X has a higher resource cost than project Y, the result may be that the combined costs of risk and resources is given by $\bar{X}$ and $\bar{Y}$ respectively. If so, in this hypothetical example, the previous ranking of X over Y on net benefit terms is reversed.

Of more interest are those costs that could be incurred in an effort to reduce the risks of health damage. Typically, the higher the expected value of damage the *less* costly it is to reduce that damage by one unit. An abatement cost curve would therefore go from left to right with negative slope in figure 4. Moreover, the abatement curve relevant to project X is almost certainly not going to be the same as the abatement curve relevant to project Y. What happens here is that a *separate* cost–benefit analysis has to be done in which the cost of reducing risk is compared with the reduction in the cost of risk (such a reduction now being a benefit).

Suppose, in both cases, such a study is done and net benefits are demonstrated. Then the effect is shown in figure 5. Note that anything can happen, depending on the initial conditions (the position of the existing total cost curves) and what net benefits are demonstrated from the risk-reducing programme. In essence, however, the total cost curve shifts downward for each project, the extent of that shift being determined by what the abatement programme can achieve. Offsetting this, but not fully, by assumption, will be the fact that costs of abatement are now incurred. The *net* effect is thus to lower the cost curve for X and the cost curve for Y *and* to lower the risk levels of each project. Which one now has the biggest net benefits *overall* depends absolutely on the empirical magnitudes resulting from the introduction of the various measures.

Conclusions and observations

I set out with deliberately limited aims and with an array of explicit or implicit assumptions, which time precludes me from elaborating. The results are, however, readily summarized. Risk analysis has become popular and it has generated a literature all of its own, and justly so. Despite the simplifying assumptions that I have made, risk is a complex and formidable problem to analyse. My aim has not been to capture much of that complexity but rather to ask the question: what do we do with risk analysis if we succeed in getting some answers? My argument has been that we can integrate it into other forms of decision-making calculus, of which cost–benefit seems to me to be the most appealing, regardless of the criticisms advanced against it. Yet once this is done, risk analysis is seen to be part and parcel of something that economists have been doing for years without feeling the need to separate it and give it special status or title. More important, I have argued that risk analysis is positively misleading in giving rise to false expectations about the role that it can play in decision-making. Possibly no one argues for it as a ranking device on its own, yet the danger is there that that is how it will be used.

I have argued that risk analysis will give the same answer as cost–benefit analysis only if some very highly restrictive assumptions are made. The salient ones are as follows.

(i) The magnitude of the event must be the same in the events being compared, and magnitude must be construed to include the 'quality' of the event. 'Risk' becomes very much a 'fuzzy' concept.

(ii) Once the unrealistic assumption of 'equalized' magnitudes (e.g. deaths per year) is dropped, we require knowledge of the sufferer's utility function and a rule for aggregating those effects. Yet once this is recognized we are obliged not to use objective probabilities but perceptions of risk as at least a two dimensional concept. Comparisons of objective probabilities of death or morbidity have little or no policy relevance.

(iii) All other costs must be equal for us to be able to assume that benefits are equal. To put it another way, one unit of electricity from wave power need not be at all the same as one unit of electricity from nuclear power. If risk analysis has to be pursued, one is tempted to ask for risk per unit of money value of benefit secured.

(iv) The idea of comparing risks between activities takes place in a vacuum, since it fails to observe that there are mechanisms for reducing those risks, but at a cost. We have seen with purely hypothetical examples that the final ranking of projects has everything to do with a proper comparison of *all* the costs and benefits, including costs of risk reduction.

To be wholly realistic, one must place risk analysis in its political context. It is there for the purpose of finding one more argument to defend this industry or that against its opponents, whether they favour nuclear power or solar power. It would therefore be naïve to expect it to go away. The least that can be done, perhaps, is to prevent it from entering the realms of high academic fashion beyond the levels already achieved in a remarkably short time.

References (Pearce)

Arrow, K. 1963 *Social choice and individual values*. New York: Wiley.

Fischoff, B. 1979 In *Energy risk management* (ed. W. D. Rowe & G. T. Goodman), pp. 269–283. London: Academic Press.

Green, H. A. J. 1976 *Consumer theory*. Basingstoke: Macmillan.

Henderson, P. D. 1977 *Oxford econ. Pap.*, July

Holdren, J. *et al.* 1979 *Risk of renewable energy sources: a critique of the Inhaber Report.* Berkeley, California: Energy and Resources Group, University of California.

Inhaber, H. 1978 *Risk of energy production*, 3rd edn. Ottawa: Atomic Energy Control Board.

Mooney, G. 1977 *The valuation of human life*. London: Macmillan Press.

Mueller, D. C. 1979 *Public choice*. Cambridge University Press.

Nash, C., Stanley, J. & Pearce, D. W. 1975 *Scott. J. polit. Econ.* **22** (2), 121–134.

Rothschild, Lord 1978 *Listener*, 30 November, pp. 715–718.

Rowley, C. & Peacock, A. 1975 *Welfare economics: a liberal restatement*. London: Martin Robertson.

Ryan, W. J. & Pearce, D. W. 1977 *Price theory*. Basingstoke: Macmillan Press.

Schaeffer, R. E. 1978 *What are we talking about when we talk about 'risk'? (Int. Inst. for Applied Systems Analysis, Research Mem. no.* RM-78-69). Laxenburg, Austria: I.I.A.S.A.

Starr, C. 1969 *Science, N.Y.* **165**, 1232–1238.

Starr, C. 1979 In *Directions in energy policy* (ed. B. Kursunoglu & A. Perlmutter), pp. 329–342. Cambridge, Mass.: Ballinger.

Starr, C., Rudman, R. & Whipple, C. 1976 *A. Rev. Energy* **1**, 36–51.

Discussion

S. Russell (*Technology Policy Unit, University of Aston, Birmingham, U.K.*). The $64000 question which Professor Pearce has glossed over – or, if we believe the American National Standards Institute, the $300000 question – is how to assess the cost of a human life. If, for example, we base it on lost production, we come up against what has been called the 'shoot-the-pensioners' paradox: they contribute nothing and should therefore be disposed of. Assigning a monetary cost to health damage can lead to dangerous corollaries. A study of energy systems (U.S. A.E.C. 1974 *Comparative risk–cost–benefit study of alternative sources of electrical energy*, WASH-1224), which attempted to use the A.N.S.I. figure in an all-embracing cost–benefit analysis – fortunately the authors had to admit its impracticability for most external costs – found that the health costs were negligible compared with the conventional costs. Are we then to ignore such health risks because our accounting system deems them insignificant?

D. W. Pearce. I fear that Mr Russell has not understood some of the basic facts that relate to living in a world of finite resources. If, as economists have repeatedly pointed out for the last two centuries, our wants exceed our capacities to meet them, we must, when choosing one thing, give something else up. This, very simply, is the notion of opportunity cost and an inescapable fact of life. Since no society devotes its entire resources to the elimination of the risk of death (if it did, we would spend our entire gross national product, bar something to enable us to survive, on the national and private health services) it follows that *society itself* places a finite value on human life. What the cost–benefit analyst seeks is a methodology for 'revealing' that underlying valuation. By referring to the 'foregone output' approach, Mr Russell is indicating an unawareness of the vast critical literature by economists who reject that approach as having little or nothing to do with how society values a life. The 'shoot the pensioner' paradox is the result of spurious economics, not opportunity cost logic. While not saying so in so many words, Mr Russell's problem is that he does not want to accept revealed preferences for the valuation of reducing the risk of death unless they coincide with his (the 'value of life' is slightly misleading since we do not speak of such valuations outside of the context of risk). If the 'accounting system' is one based on the individual's values, I suggest there are some good reasons for accepting the results, and the strongest of these is that the valuations relate back to what people want. If Mr Russell's argument is that we should override those values because 'we' somehow know better, then he faces the classic paradox of liberalism: when and when not to reject society's valuations in favour of some élite's 'better judgement'. As it happens, cost–benefit analysis is perfectly capable of embodying élitist values. What it rejects, and rightly so, is any notion that human life is either infinitely valued or 'beyond' value. Both notions are entirely inconsistent with how societies behave. I see far more danger in an unwillingness or incapacity to follow opportunity cost logic through to its conclusions than I do in the attempts made so far to place money values on risk reductions as they affect life and death.

R. F. Griffiths (*Pollution Research Unit, U.M.I.S.T., Manchester, U.K.*). Professor Pearce has criticized Dr Inhaber for expressing death in terms of man-days lost, and I would agree with such criticism; Professor Pearce has himself been criticized for encouraging us to express death in terms of money. Would it not be best to leave the impact expressed in the form in which it actually arises, rather than trying to produce a unified index, which is a process on which we are never likely to find agreement?

D. W. PEARCE. My response to Mr Russell explains the logic of placing money values of risk changes as they affect human life. I do not therefore accept the criticism that economists are at fault in attempting such valuations, whereas I do maintain that Dr Inhaber's exercise is very largely without meaning. There is, however, an added reason for not leaving 'human life' as an unquantified item in a cost–benefit analysis. Once we abandon the search for common measures of social value, we find ourselves with something akin to environmental impact matrices in which dozens of attributes are given dozens of scales of value, few of them comparable. What follows from such a situation is that unless all of the attributes in one matrix are clearly superior or inferior to all of the attributes in another, we have no basis for choice. Such inequality of attributes is not only rare but actually explains the very emergence of cost–benefit analysis as a decision-aiding procedure. Finally, once we do leave an attribute unvalued we come full circle to valuing it when the decision is taken! To choose a project costing £X million and which risks 10 lives over a project which costs £2X million and risks no lives is to value the 10 lives at a *maximum* of £X million. If we choose the latter project we are saying the lives are worth a *minimum* of £X million. Like it or not, society does value human life in finite money terms.

Proc. R. Soc. Lond. A **376**, 193–197 (1981)
Printed in Great Britain

An industrialist's attitude to risk

By P. G. Harvey
Imperial Chemical Industries Limited, Imperial Chemical House, Millbank, London SW1P 3JF, U.K.

The definition of risk is considered and the distinction made between risk and hazard. Some risks are unacceptable because the potential for disaster is so great or because the probability of occurrence is not sufficiently definable. It is argued that acceptance of residual risk is implicit in all our activities and explicit in the U.K. Health and Safety at Work Act. Ways in which criteria of acceptability might be developed are discussed, bearing in mind that the issue will not always be resolvable by numerate analysis. In looking at the way ahead for industry, the author accepts the need for regulation but pleads that society should join with industry in defining criteria of acceptability so that technological progress is not frustrated or even arrested.

I must state at the outset that the views expressed in this paper are those of an industrialist influenced by a scientific education and a lifetime in science-based industry. Thus wherever possible I look to numerate analysis as a major contribution to problem solving, whereas others may not be so inclined, particularly in the field of safety. However, we doubtless have common aims, namely the containment of risk within reasonable limits, and I have framed my paper with this in mind.

I well recall my own reaction as an inexperienced but enthusiastic young plant manager some 30 years ago to repeated exhortation to improve safety performance from senior mangement and directors. Too often, it appeared to me, the exhortation demanded the elimination of risk, and since the technology that we were managing was intrinsically dangerous it appeared the only sensible response to the exhortation was to stop manufacturing lethal gases. When I was a works manager, a later generation of plant managers asked me which came first, safety or production; I was not entirely comfortable with my answer. I therefore welcome the introduction over the last decade of quantification of safety assessment and believe that it has raised the whole credibility of safety and accident prevention in the eyes of our production management.

Having allowed the scientist in me an opening gambit I must remember that you are expecting to hear from the industrialist, who may or may not be a scientist. The industrialist is used to taking all kinds of risk, all of which he likes to think are calculated. His business is about risk taking and getting the right answers to assure the prosperity of the enterprise. Safety is but one of the risks, albeit an important one, to be assessed and weighed with the other uncertainties in the whole venture.

Focusing now on this question of risk to people that will be presented by particular industrial activities, we in the chemical industry have by the nature of the business our fair share of major hazards. Despite the occurrence of a disaster such as that at Flixborough, our record in this country is good. Indeed, based on this record it has been calculated that, if all other sources of death were removed, including old age, and the only cause of death was accidents in the chemical industry, we would all live for an average of 12500 years. This compares with 1300 years if we drink a bottle of wine a day and only 100 years if we smoke 40 cigarettes a day. Thus it would seem that working in the chemical industry ranks among the most risk-free of human activities. This is not say that there is room for complacency, for clearly the potential for doing harm exists.

This leads me to a *definition of risk* that appeals to me and comes from the *Second report of the Advisory Committee on Major Hazards*; it is: 'The probability that a hazard may be realised at any specified level in a given span of time.' I am not arguing at this stage that any particular level of risk should be acceptable for a given hazard, but I am saying that we must be clear on our points of argument. All too frequently, hazard – i.e. the *possibility* of a harmful event – is confused with the *probability* that the potential of the hazard will in practice be realized. This popular misconception is at the root of at least some of the disquiet over, for example, the use of nuclear energy for peaceful purposes, although I do accept that, however low the mathematical probability of a man-made disaster, if its consequences are very great, people may still not be willing to accept the risk. I shall return to this point later, but meanwhile let me say that it is my belief that the vast majority of industrial activities are controllable to such a degree that, once the hazards have been identified, the risk, as defined above, can be made acceptably low.

I now turn to the identification of hazards and the ways in which they can be controlled. Systematic hazard study techniques that are in use in industry are applied at the various stages of a new venture, starting with the inception and proceeding through design and construction to the operating stage. Essential parts of these studies are both a check against past experience in similar activities and the anticipation of new hazards by detailed critical examination of the elements of the project. Many of the hazards identified in these studies will be of a kind that will be adequately controlled by adherence to statutory requirements, codes of practice and other in-house safety measures that have been proven with comparable hazards. In ventures that contain novel features, such as new chemicals, advanced technology or increases in scale, there will of course be areas of doubt. In these areas, judgements will need to be made on the special safeguards required, or sometimes on the need to find a safer route.

This brings me back to the *unacceptability of some risks*, either because the hazard is judged to be unacceptable to society or because the risk is found on examination not to be sufficiently controllable. In the former, I recognize as an industrialist that society is more tolerant of the Lord than it is of me. Thus Californians will continue to live happily on the San Andreas fault but protest vehemently against an unnatural risk arising from a chemical or nuclear venture. In some such cases I would bow to the collective wisdom of man and accept that the venture was not viable.

Similarly it sometimes happens, albeit rarely, that a venture exhibits risks that at the current state of knowledge are poorly definable or would be too costly to reduce to an acceptable level: again the venture would have to be abandoned. But these are rare situations: the norm is where risks of so-called 'major hazards' can be managed at tolerable levels.

Implicit in what I have just said is that *residual risk* does and always will, exist in industrial ventures, just as it does in all our everyday activities. It is to the credit of the legislators in the U.K., unlike those in some other countries, that this fact is recognized by statute. The Health and Safety at Work, etc., Act, in calling for various safety measures, carries the proviso '...so far as is reasonably practicable', and this is surely a refreshing air of realism.

I dwell on this acceptance of residual risk because, as with the confusion over 'hazard' and 'risk' already mentioned, one does detect a mental block on this subject in some otherwise reasonable people. This is particularly, though not exclusively, true when the risk has to do with the more hazardous installations. I suspect that the problem arises because of our difficulty as humans in accepting that some bodily harm is inevitable, indeed almost planned, as a result of our daily work. Man's inhumanity to man must it seems be restricted to outright war, pitched battles on the football terraces, or the 'sport on the roads'.

Having emphasized the reality of residual risk in industrial ventures, I cannot now escape some discussion of the *criteria of acceptibility*. And here I must draw a distinction between what is socially acceptable and the criteria, or yardsticks, that industry needs for what I shall loosely describe as 'design purposes'.

Social acceptability cannot be defined solely in numerate terms. The public apply qualitative judgements to certain risks; radioactivity and carcinogenicity are two obvious examples where society demands higher standards than for, say, violent deaths on the roads. Furthermore, there is an understandable prejudice in favour of higher standards for occupational risks that are not as seen as voluntary. Thus, I accept that it is no argument in dealing with occupational carcinogenic risk to equate this risk to smoking, which is perceived as an individual voluntary choice.

Industry, however, does need yardsticks by which to judge the acceptability of its processes, and while the continuing debate, as exemplified by the present proceedings, is fascinating, life does go on and targets have to be set. In fact, for more than a decade, during which time risk quantification techniques have been under development in industry as an aid to decision taking, it has been necessary to fix levels of acceptability appropriate to the ventures. In my own company we have aimed to improve on past performance by using criteria based on historical accident frequency data. Statistics for the past decade show a significant reduction in process-related fatalities compared with the previous decade, and we would claim that this success is at least in part due to the use of the new techniques. In passing, it is perhaps worth noting that this method of considering appropriate criteria (namely by an analysis of past experience) has been possible only for people at risk within the factory fence, for the simple reason that we have, to date, substantially contained the effects of our incidents within our fences. It is largely an unrecognized fact that, as far as I can determine, no member of the outside public in the U.K. has been killed as a direct result of an accident on our process plants.

I must emphasize that the analytical techniques to which I have referred are not in themselves decision-making tools. This would appear to be contrary to what is imagined by some other speakers at this conference, but that is not my experience. Nor do I foresee that the point will ever be reached where a purely mechanistic approach to risk taking can be adopted: the final decision will always be to a large degree judgemental.

And so, what of the future? It has been said elsewhere that this is the first generation that feels itself to be entitled to immortality. Somehow I feel that this is an overstatement, but I do recognize that society is becoming ever more demanding, and this is understandable. However, I must say to society that we must be careful to proceed at a pace that we can afford.

I accept that some regulation by statute is needed for the major hazard plants, to ensure that industry gives due attention to the risks involved. But such regulation must be of a kind that allows industry the flexibility that it needs to take the technical decisions needed to achieve a reasonable level of risk. I must stress, even though it might sound arrogant, that industry alone has the competence to make the right technical decisions, and this fact needs to be more widely appreciated. I remain convinced that self-regulation to high standards is the greatest safeguard to employees and the public alike. I am not ruling out the need for external monitoring by the Health and Safety Executive, provided this is not unnecessarily bureaucratic. In fact, the proposed U.K. Hazardous Installations (Notification and Survey) Regulations, currently being developed in consultation with industry, could well fill this gap admirably.

However, this does leave open the question of what is a reasonable level of risk. The answer will remain obscure until ways can be found of equating society's high ideals with what is practicable and economically possible. Clearly a partnership is needed to find these ways, and somehow or other we must come together.

I conclude by reminding you that private business is itself about risks of all kinds. Any new venture has associated with it a range of hazards, one of which is the safety of the enterprise, which has to be weighed along with the other uncertainties. Society still expects technological progress; although it is true that this should not be at *any price* in *human* suffering, there is some such price that it must be willing to pay.

Discussion

D. R. Cope (*Energy and Planning Group, Institute of Planning Studies, University of Nottingham, U.K.*). Despite the hopes expressed by the chairman of this session, there has been little discussion of the political dimension of risk assessment and hazard management. A contribution from a politician might have illuminated the lack of congruence between 'objective' and 'subjective' assessment commented on by several speakers.

We have spent two days discussing risky events and the consequences that follow them. There has been little recognition that there are many consequences just waiting for risky events to occur. It is, for example, well known to many local pressure groups that the swiftest way to ensure installation of a pedestrian crossing on a dangerous road is for a fatal accident to occur there. This may have more effect than any amount of preceding formal political activity.

It is helpful to identify for each risk the 'risk constituency', that group of the population that is at risk from a particular hazard and who might be expected to take political action to minimize such risk. In some cases such constituencies are easily recognizable and well organized. Trade unions are an example and this helps to explain the origins and emphases of the Health and Safety Executive. Other constituencies are more amorphous and less well organized, although they may rapidly gain political influence after the occurrence of an event. There may therefore be a paradox that it is in the collective, although obviously not individual, interest of such constituencies for risky events to occur for the reasons mentioned.

My second point concerns the timing of hazardous events and their relation to the ongoing process of assessment and policy-making in broad strategic policy areas such as energy supply policies. Most Western countries are currently engaged in such exercises. The niceties of the 1 in n event, where n is a very large number, are lost even to supposedly objective investigations. For example, it is probable that the events of Three Mile Island will exercise a disproportionate influence on the forthcoming public inquiry into the Sizewell p.w.r. to the detriment of consideration of other risk possibilities. Similarly, I feel that the agenda and outcome of the recent Vale of Belvoir coalfield inquiry would have been very different had a shaft accident such as that at Markham Colliery occurred during or immediately before the inquiry.

In any future discussions of risk assessment and hazard management it would be useful to give attention to how issues reach the political agenda and public consciousness and how they fare once there.

Proc. R. Soc. Lond. A **376**, 199–204 (1981)
Printed in Great Britain

Regulation of risk

By H. J. Dunster

Health and Safety Executive, Regina House, 259 *Old Marylebone Road, London NW*1 5*RR, U.K.*

Health and safety regulations have always been concerned with risk, though not always overtly. The quantitative expression of risk does not appear in regulations and rarely in guidance materials but is inherent in the policy underlying the development of regulations and in their practical application.

The ways in which actual and perceived expressions of risk are used in regulatory actions differ widely. Some dangers are treated as unacceptable and the regulatory policy is to exclude them totally. In the real world, such policies are never completely successful. The head-on collision of trains in main-line working is one example. The deliberate use of known carcinogens as pesticides is another. Other dangers are recognized as inevitable but as being reducible in degree. The regulatory activity is then aimed at limiting the extent to which a citizen can expose other citizens to this danger and, more recently, the extent to which he is permitted to put himself at risk. The balance of risks and benefits and of one risk with another underlie decisions in these cases. Not only consequences but probabilities become relevant. In the past, all, and even now most, of the regulation of risk has been on a non-quantitative basis. Increasingly, there is a desire to make the process more quantitative and to introduce the idea of acceptability. This change is provided for by many of the features of the Health and Safety at Work Act of 1974. Within the framework of that Act, the Health and Safety Commission and its operating arm, the Health and Safety Executive, are developing the more systematic use of quantitative methods of controlling hazards from work activities.

The regulation of risk is a growth industry and it behoves us all to clarify our objectives. An aim of zero risk would not be to the benefit of society, but its replacement by more suitable aims is a long and complicated process.

Introduction

The word 'risk' is not consistently used or defined. It sometimes means the probability of an undesired event, as in the risk of drowning, and sometimes the events themselves, as in the risks of rock climbing. Sometimes it is an indefinable mixture of these two. In this paper, it will be used to mean a qualitative combination of the probability and severity of an undesired event.

Health and safety regulations have always been concerned with risk, though not always overtly. The quantitative expression of risk does not appear in regulations and rarely in guidance materials but is inherent in the policy underlying the development of regulations and in their practical application.

Health and safety regulations related to work activities in Britain are aimed at eliminating dangers to health and preventing injuries and death. Such an absolute objective has to be tempered by reality if it is to be effective. Some regulations are indeed absolute, e.g. the requirement to guard moving machinery, and require specific and feasible action. Even complete compliance with these requirements does not totally eliminate the danger. A guard or its interlock system may fail. In the past, some regulations have been absolute in requiring specific action that was not feasible. The 1931 Asbestos Industry Regulations required that there should be no asbestos dust in the workplace. Such requirements are neither equitable nor effective.

Other regulatory requirements have been less specific and have contained qualifications, culminating in the general duties placed on employers by the Health and Safety at Work, etc. Act of 1974. These duties, in effect, put an onus on the employer to remove risks, but the onus is qualified by the phrase 'so far as is reasonably practicable'. Underlying this formulation is a clear recognition that risks exist and will continue to exist, but there is also an identifiable underlying policy that says that risks should be reduced and that the degree of effort in reducing them should be tailored to the magnitude of the risk. Where risks are high, absolute requirements still have a place and substantial efforts must be made to implement non-absolute requirements. Where risks are low, the onus is less.

A basic regulatory policy

By considering the existing regulatory approach to risks, it is possible to distinguish two distinct policies. In one, the aim is to reduce the risk to zero unless something goes wrong. In the second, the aim is to establish a normal situation in which some risk is still possible and to keep this risk as low as is reasonably practicable. Neither of these approaches leads to a total absence of risk because, even in the first case, there is a finite probability that something will go wrong. In seeking for a consistent policy for the regulation of risk it is extremely important to recognize the difference between these two cases and to choose a solution which is both appropriate and possible. Attempts to use the first policy when the second is appropriate leads to inconsistent and unenforceable regulations.

Some practical examples

In practice, the pattern of regulation is influenced by industrial developments and by public and political pressures. The simple idea that the phrase 'so far as is reasonably practicable' links the scale of effort with the scale of risk is heavily modified by people's differing perceptions of risk. In assessing the severity of a risk it is necessary to consider the likelihood, the consequences, and also the perception of those factors by those who are or may be exposed to the risk, or by those who represent the exposed group. More recently, pressures have also been exerted by those who seek to speak for society in general. The influence of the different perceptions changes the priorities of employers, employees and regulatory agencies. They can no longer seek, even in the ideal, to minimize harm. They must now move towards the aim of minimizing the perceived harm. There is, I suggest, an implied duty on all of us to help keep the perception of harm broadly in line with the magnitude of that harm. If perceptions are seriously out of line with magnitudes, all our priorities will be distorted. Complete alignment is neither feasible nor, indeed, desirable.

Railway safety

The aim, successfully achieved, of the railways and those who regulate them is to make it possible to regard railways as a safe method of travel. The possibility that a passenger will not arrive safely at his destination simply does not enter his head. When this perception is shaken, as by major accident, public opinion expects the cause to be eliminated and that is also the intention of the Railway Inspectorate. Even in this apparently absolute situation there are shades of emphasis. Head-on collisions in mainline working are virtually non-existent, except at the crossovers associated with junctions. The residual risk is too small to justify flyovers at these junctions. Head-to-tail collisions do still occur, sometimes due to the failure of signalling systems. Economics have some part in limiting the reliability of these systems and there are also practical limits because the extensive application of fail-safe systems would mean that the systems had to be overridden after failure, to keep trains running – a potentially more dangerous situation.

The safety of pesticides

The aim of the Pesticides Safety Precautions Scheme is said to be to exclude harm entirely for humans, provided that the product is properly used. The logic of this approach is perhaps open to question: improper use will certainly occur and cause some risks. The pursuit of zero risk from proper use may then involve some misuse of resources.

One development of the policy relates to pesticide residues in foodstuffs. If a new pesticide were to be assessed as a carcinogen in man and would persist in any concentration in foodstuffs, its use would not now be approved. Implicitly, it is assumed that a carcinogen does not have a threshold below which there is no risk. While existing pesticides are adequate and new ones not dramatically better, this policy is sound in its context. Even so, it has one serious drawback: it puts great emphasis on the definition of carcinogenicity. It shares this disadvantage with any regulatory scheme that treats carcinogens by a different regulatory régime from that for other toxic materials.

The safety of chemicals in the workplace

Early toxicity studies tended to concentrate on the observable effects following acute or short-term chronic exposure. If a level of exposure could be found below which most people, or in the case of experiments, most animals, showed no effect and the incidence of severe or irreversible effects, such as malignancy or sensitization, was too small to observe, that level, or more usually some fraction of it, was called a limit, eventually a threshold limit value. More recently, the possibility of severe effects at low probability after long continued low exposures has led to fears that some, or perhaps many, substances have no true threshold of effect. In the regulatory sense, compliance with a threshold limit value, or other industrial hygiene standard, for a material confidently believed to have no adverse effects at lower levels of exposure, would be legally sufficient, although in view of the uncertainties of toxicology, not necessarily prudent. This position would, however, be changed if there were scientifically plausible forecasts of low-level harmful effects, even if these had not been directly observed in man. Thus a material known to be carcinogenic in mammals at levels not grossly above those relevant to human exposures might reasonably be foreseen to be carcinogenic in man. When carcinogenic materials have their primary influence on the organ of entry, the lung, the gastrointestinal tract or the skin, it may well be reasonable to believe that there is no identifiable threshold and that there will be some residual risk, even when exposures are very low indeed. Asbestos and ionizing radiation are two well documented examples of agents that can act directly on organs or tissues of the body. Where a chemical has to pass through metabolic processes to reach the organs where it is likely to have a carcinogenic effect, then the possibility of a threshold is much more real.

For materials with a genuine threshold and for which a limit has been set below that threshold, the only risk is that arising from failure to comply with a limit. In Britain, there is an onus implied by the general duties of the Health and Safety at Work Act to reduce the likelihood of such failures so far as is reasonably practicable. For materials that do not have a threshold, there is some risk at or below any prospective limit and there is then an onus to achieve further reduction of exposure. Only when the cost (in money, time or trouble) of the next step towards reducing exposures or risks of failure is out of proportion to the expected reduction in harm is the statutory duty formally discharged.

It is sometimes argued that an effective threshold might be defined for any carcinogen so that no additional cases of cancer would be discernible. Unfortunately, cancer, while fairly rare in youth and middle age, becomes increasingly common at older ages and a protection policy based on lack of detectability would be inadequate for those cancers with a high natural frequency. The approach would also be extremely sensitive to the size of the exposed group and would thus lead to severe anomalies.

A more rational policy has been identified by the Advisory Committee on Asbestos. It is a policy that can be applied only when there are adequate data about the relation between the probability of cancer and the level of exposure. The approach involves studying the actual levels of exposure across a wide range of industries dealing with a carcinogen. The parts of industry where the exposures are most difficult to control are considered first and a limit proposed that would be achievable in these areas, though only by the expenditure of substantial effort and money. The risk associated with this limit is then compared with the general range of industrial risks. If it is not seriously out of line, that limit is then taken as an overriding limit. Other parts of industry should then be able to do better and they are subject not only to the limit but also to the general requirement to keep exposures as far below the limit as they reasonably can. This two-pronged approach, which has been commonplace in the control of ionizing radiation for some decades, now appears to be the most appropriate technique for dealing with any situation where compliance with a stated policy or regulation cannot be expected to reduce the risk to zero.

The technique cannot always be applied quantitatively. Often the biomedical data are inadequate and information about levels of exposure in industry not yet forthcoming. The technique then becomes semi-quantitative or completely qualitative. It remains a useful form of self-discipline and ensures that materials with poorly understood dangers are treated with respect. Quite simple precautions may provide protection, even against unforeseen hazards, and then should not be rejected merely because the need cannot be clearly defined.

Clearly, these are not matters of science alone. When scientists participate in the final decision they are acting as administrators, regulators or politicians. They need to remember the distinction.

Future developments

It is not easy to see a way forward to a consistent regulatory approach to risks. In some countries, notably the United States of America, the basic legislation still prevents a consistent approach and the various Government agencies, including the regulatory agencies, are still feeling their way forward. There is, as a result, some considerable confusion in Europe about American intentions. It is difficult on this side of the Atlantic to be sure whether a document represents a proposal, a statement of agency policy, an agency policy that has been confirmed by a Court, or a decision of Congress. The public issue of advice from one agency to another, when that advice may or may not be accepted and, if accepted, may be subsequently overturned, can have dire consequences for countries who hope to draw on American experience as an example. But the absence of consistency is itself serious. Unexplained inconsistency casts doubts on the validity of all of the propositions. The blame is often put on the experts, where it should really lie on the regulators or legislators. In the international context, inconsistency leads to non-tariff barriers to trade – a very undesirable feature in the eyes of the European Economic Community.

Complete consistency is an unrealistic aim in view of the different national legislative structures. We should, however, be aiming to increase the amount of consistency, and perhaps the best way of achieving this is by continued discussions of the kind stimulated by this meeting. Organizations such as the European Economic Community and the Organization for Economic Cooperation and Development have important roles to play but must recognize that consistency can be bought at too high a price. I would rather see us inconsistent and right some of the time than completely consistent and always wrong.

Discussion

G. H. Kinchin (*Safety and Reliability Directorate, Culcheth, Warrington, U.K.*). Mr Dunster suggested that the effort should be greatest in areas of high risk. Presumably he means perceived risk, even though this will not maximize the number of lives saved.

H. J. Dunster. There is no simple reply. Some attention has to be given to the perception of risk but this must not totally override the objective assessment. A judgement of the necessary weighting factor has to be made by the regulatory agency.

P. J. Bunyan (*Slough Laboratory, Ministry of Agriculture, Fisheries and Food, Slough, U.K.*). I should like to take issue with a statement made by Mr Dunster during his presentation. I do not believe that anyone associated with the Pesticide Safety Precaution Scheme would claim that they achieve zero risk for the use of pesticides if the label recommendations are followed. I am certain that this would only be seen as reducing the risk to an acceptable level. I do, however, agree that the scheme has great difficulty in dealing with the question of misuse, i.e. use contrary to the label recommendations. Professor Farmer had suggested that in other situations, safeguards could be engineered into plant to override the human factor. For individual workers on farms, who may be using biologically active chemicals, no such safeguards can be built in. Would Mr Dunster comment on his approach to this problem?

H. J. Dunster. My understanding of a statement of the Chairman of the Advisory Committee on Pesticides was that the *aim* was zero risk, given proper use. Perhaps I misunderstood him. In either situation, some level of risk has in fact to be accepted and my main point is that this is not a matter of science alone. A committee of experts alone can advise on the likely level of risk, but if they advise on the acceptability of that risk they are no longer advising as experts.

Proc. R. Soc. Lond. A **376**, 205–206 (1981)
Printed in Great Britain

Concluding remarks

By Sir Frederick Warner, F.R.S.
Messrs Cremer and Warner, 140 *Buckingham Palace Road, London SW*1*W* 9*SQ, U.K.*

As I flew last week over the Mekong River, then the Ganges Delta in Bangladesh and on over Gujarat, I could not help reflecting on the scale of catastrophes, from the 1931 estimate of two million deaths in one flood in China to the 15000 in the 1979 collapse of a dam in Gujarat. They have been listed by Fryer & Griffiths (1979). The ability to kill many people in man-made accidents is limited by the containment, whether in a car, train, aircraft or factory. Containment possibly makes society tolerate these risks while questioning the smaller risks, made all-pervasive by the concept of dose-commitment in radiological protection. When the same concept is used by Dr Inhaber to predict risk from SO_2 and airborne fluorides, many part company with him, although accepting the general thesis that economic activity carries known risks and spending money brings death or injury.

This Discussion Meeting has ranged widely and has shown up some gaps. It has especially focused attention on the differing perceptions of risk and the failure of quantitative risk assessment to make much impact on them. Lord Ashby suggested that insurers know how to assess risk from past experience and sell policies by judging the perception of that risk. On the whole, they do not offer better terms than 10^{-4}/year in familar areas like house insurance. When it comes to betting, a 100 to 1 outsider attracts little money, and most of the figures discussed here mean nothing to the average man. They tend to agree with Damon Runyon's assessment that everything in life is 6 to 4 against.

Professor Pearce has told us that these questions have been settled for a long time by economists, and cost–benefit analysis can put a value on human life. This does not help when the values differ from Sir Edward Pochin's $15 for saving a death from smallpox to the $20M per life from higher building standards after Sir Alfred Pugsley's report on Ronan Point. Deaths are not seen as equal; multiple deaths require a power function until they reach the stage of catastrophes, when the perception rapidly decreases. Nor has there been much light thrown on fates worse than death.

The small-scale dramatic accident and the unfamiliar are given great weight in industrial societies, which have the need for official inquiries to attach blame and find a scapegoat for the failure of organizations to recognize a risk. The Flixborough disaster, for example, led to an extended investigation, and the rebuilt plant has a process, involving phenol hydrogenation, that gives overall hazards greater than the cyclohexane oxidation process that it replaced.

As for the time effect, the recent release of 40 Gy at Three Mile Island has created great anxiety, whereas the experience of burning 40 t of highly radioactive rods in a stream of air during the Windscale accident in 1957 has now no impact. Professor Okrent has shown this in comparing proposals for l.n.g. storage with the acceptance of l.p.g. storage in built-up areas. Of longer standing is the location of holders for gas in town centres. He emphasized the social and political problems in drafting safety laws.

Dr Thomas described the use of surveys in Austria to measure public perception of nuclear power, but work in Sweden has shown that perception can change greatly after, and perhaps as a result of, the survey. Professor Lee puts hazard back into the area of moral and ethical judgement, providing little comfort for those whose conscience leads them deliberately to give priority to those areas that their work has shown to give large and measurable saving and extension of life.

On behalf of the Royal Society I express thanks to all the authors and those who have contributed to the discussion.

Reference (Warner)

Fryer, L. S. & Griffiths, R. F. 1979 *World wide data on the incidence of multiple fatality accidents*. SRD R149. London: H.M.S.O.